PORTRAIT *of a* WOODLAND

BIODIVERSITY IN 40 ACRES

Dedication

For my sons Nicholas and Martin and especially my grandchildren Michael, Benedict and Charlotte who I hope will get as much interest, fascination and enjoyment from the woodland as I have, and who, I hope, will see the woods in all their grandeur when mature.

By the same author and published by Search Press Ltd:
Plant a Natural Woodland – A Handbook of Native Trees and Shrubs
Culinary Herbs and Spices

As Charlotte Green:
Gardening Without Water
All About Compost

Acknowledgements

Without the help of the organisations listed below the making of this book would not have been possible. The individuals concerned are named in the text and/or on the acknowledgements page.
The views expressed in this book are my own and not necessarily those of the organisations.

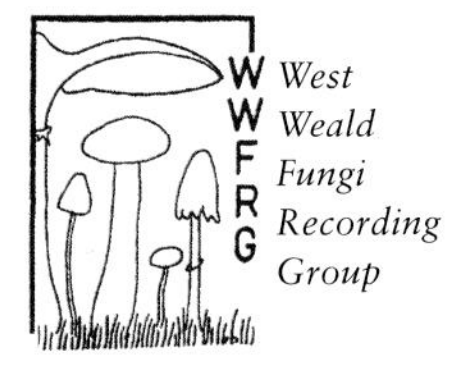

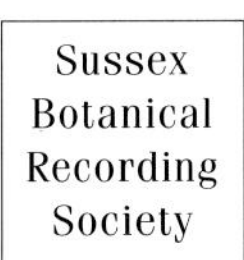

When I behold the havoc and the spoil
Which, even within the compass of my days,
Is made through every quarter of this isle,
In woods and groves, which were this kingdom's praise.

George Wither (1588–1667)
written in 1635

To see a world in a grain of sand
And a heaven in a wild flower
Hold infinity in the palm of your hand
And eternity in an hour.

William Blake (1757–1827)

What would the world be, once bereft
Of wet and wildness? Let them be left,
O let them be left, wildness and wet;
Long live the weeds and the wilderness yet.

Gerard Manley Hopkins (1844–1889)
written in 1881

PORTRAIT *of a* WOODLAND

BIODIVERSITY IN 40 ACRES

Foreword by David Bellamy

CHARLOTTE DE LA BÉDOYÈRE

SEARCH PRESS

First published in hardback in Great Britain 2004
Search Press Limited
Wellwood, North Farm Road,
Tunbridge Wells, Kent TN2 3DR

ACKNOWLEDGEMENTS

Grateful thanks to my editor, Tim Harris, for his sensitive editing, and to Bruce Middleton for much needed advice which has kept me from too many blunders or howlers.

Special thanks to the following individuals (in alphabetical order) who helped to piece together the biodiversity in the forty acres of woodland, and who patiently put up with my endless queries and phone calls:

Henri Brocklebank
Sussex Biodiversity Record Centre
Woods Mill, Henfield
West Sussex BN5 9SD
Tel: 01273 497553 / 554
Email: sxbrc@sussexwt.org.uk

Janet E Claydon and Barry Kemp
Sussex Amphibian and Reptile Group
9 Kingsway, Seaford
East Sussex BN25 2NE
Tel: 01323 492066

Richard Everett
Forestry Commission
Park Lane, Goudhurst
Cranbrook, Kent TN17 2SL
Tel: 01580 21123
Email: Richard.everett@forestry.gsi.gov.uk
Email: FC.seeng.conse@forestry.gsi.gov.uk

Chris Gent
Moor Green Lakes Nature Reserve
20 Towers Drive, Crowthorne
Berkshire RG45 7OR
Tel: 01344 773802

Paul Harmes
Sussex Botanical Recording Society
10 Hill Croft, White Oak Road
Portslade, East Sussex BN4 2QD
Tel: 01273 880258

Roger Jones and Caroline Stone
Sussex Bat Group
Email: surveys@mountfield.supanet.com

Malcom McFarlane
British Bryological Society (Sussex Branch)
Meadow View, Chapel Lane
Blackboys, Uckfield
East Sussex TN22 5BL
Tel: 01825 890125

Jacqui Middleton
Sussex Lichen Recording Group
Email:
Jacqui@brucemiddleton.freeserve.co.uk

Colin Pratt, F.R.E.S.
Sussex Moth Group
5 View Road, Peacehaven
East Sussex BN10 8DE
Tel: 01273 586780
Email: colin.pratt@talk21.com

Peter Russell
West Weald Fungi Recording Group
15 Graham Avenue, Patcham
Brighton, East Sussex BN1 8HA
Tel: 01273 505555

Sussex Branch Butterfly Conservation
Wellbrook, High Street
Henfield, West Sussex BN5 9DD
Tel: 01273 492279

Dr. Tony Whitbread and Staff
Sussex Wildlife Trust
Woods Mill, Henfield
West Sussex BN5 9SD
Tel: 01273 492630
Email: enquiries@sussexwt.org.uk

I would also like to thank the following:

East Sussex County Council Archives for permission to reproduce the maps on pages 16 and 17.

Dr. Owen Johnson of St Leonards, East Sussex for help in identifying some of the trees, especially the cloned Poplars.

Dr. Jennifer Owen of Leicester for use of the quote regarding hover-flies on page 15.

Paul Mabbott for helping me to identify the rather unusual ladybird on page 164.
www.ladybird-survey.pwp.blueyonder.co.uk

Alan Shears for the use of his photographs on pages 127, 129, 130, 131, 136, 137, 138, 139 (top), 141 (top) 142 and 143.

getmapping for use of the aerial photograph at the bottom of page 17.

Pam Watts and James Murray-Watson for making possible the wonderful helicopter trips over the woodlands that enabled me to photograph the trees and woods from a totally different perspective.

I also have to thank **Nick Moon** who has worked on maintenance throughout the forty acres and who is largely responsible for the careful clearance of the conifers in Morgan's Strip; **Dave Bates** who tirelessly pulls up unwanted Brambles and has helped to rid the forty acres of its obnoxious Rhododendrons and Laurel; **Nick Hilton** who has maintained the woodland rides for the past fifteen years
and the many others who have helped to nurture the old and new woods and maintain the many habitats. Also the many people who have given me useful information over the telephone.

Contents

Foreword

Portrait of a Woodland is the portrait of a real life dream that is still coming true. Above all, it is a very useful ready reference in words and pictures of the natural history of someone's backyard. It is the story of a little bit of heaven that caught the imagination of the author and took over her life. As the story unfolds, you will discover a lot about natural history and how Charlotte de la Bédoyère became a part of it.

Thanks to one of the great triumphs of the millennium, the much-discussed human genome project, we know that we share at least half our genes, and hence most of our living chemistry, with every plant. It has always been true to say that we are what we eat, but now we know we eat what we are. In essence, we are but part of an ongoing tapestry of the biodiversity of survival.

Let's begin at the beginning: once upon a time a lady discovered her dream family home, surrounded by trees. Further investigation showed that much of the woodland was ancient. Despite the fact that there was lots of hard work to be done she lived happily in this knowledge until the Great Storm of 1987 shattered any illusions about the permanence of trees. Until then, we had all taken them for granted.

Charlotte also learned that left to her own devices Mother Nature would do her best to heal all wounds, but with a little bit of love and expert help the back-to-nature processes could be accelerated. And so she became embroiled in naming names as she sought to discover just how many neighbours she really had in her forty-acre plot, and how best she could work with them to heal the effects of the hurricane.

Calling on the help of just about every expert natural historian in Sussex, and others from around the country, the list of Latin names grew longer and longer: trees, shrubs, herbs, ferns, mosses, liverworts, lichens, fungi, and animals of every shape and size. This amazing list is still growing, and as far as the real experts in the more microscopic things will tell you, the list of biodiversity to date is only just scratching the surface.

Yes, in some ways *Portrait of a Woodland* is a reference book that can let you into the same secrets of any scrap of woodland across the length and breadth of England, for they all share many living things in common.

Perhaps the most surprising fact that came into sharp focus was that a bit of disturbance can do the biodiversity a lot of good. So Charlotte's living web is today a patchwork of young bits, old bits and in-between bits, all alive-oh, with grassy rides, marshy areas, ancient trees, and tall straight ones growing through.

A recent survey by the Wildlife Trusts (there are 47 of them, covering the UK) showed that Charlotte is not alone. Thousands of acres, bits and pieces of every shape and size, are being given lots of love and care. Each one is a personal nature reserve. I am sure that this book will strengthen their carers' endeavour and spur others to do the same.

Take heart all those who can never aspire to 40 acres, for a suburban garden nurtured in the right way can become a biodiverse patch that will be forever England, Scotland, Wales or Ireland.

And, by the way, if you want to name names... join your local Wildlife Trust, whose experts can help you on your way.

Pendunculate Oak
Quercus robur

Brilliant sunsets can have spectacular unnatural effects on our Oaks in Autumn. During the day their leaves are normally strong browns and yellow in colour.

David Bellamy, Bedburn, March 2004

INTRODUCTION

Why write a portrait of a woodland? It is a question I have been asked many times. The reasons are almost as diverse as the number of tree species we have here. When I first purchased the house and woodlands nearly twenty-five years ago, I was so delighted and enchanted at having finally found a dream location – a house surrounded by trees and woods. I felt I ought to produce something (not necessarily a book) as a kind of legacy for my sons and their children so that they would in time become as absorbed and entranced with all that went on here. At the time I was not even too bothered by the fact that at the back of the house there existed only unrelenting conifer plantations (not yet in my ownership) that often reminded me of a quote from *Hiawatha* in which I had performed as a child:

> *'Dark behind it rose the forest,*
> *Rose the black and gloomy pine-trees,*
> *Rose the firs with cones upon them:*
> *Bright before it beat the water...'*

With a bit of imagination, the water could well be the large pond near the house. But I did not find the time to embark on this kind of project – I was too concerned about rescuing some of the garden, which was in danger of being taken over by brambles, thistles, bindweed and the like, and I wanted to create an organic vegetable garden.

It was not until the great hurricane of 1987 (which I shall refer to as 'The Storm') that I really began to think in terms of a book. However, what I actually wrote was nothing like this one, but *Plant a Natural Woodland – A Handbook of Native Trees and Shrubs*. The traumatic experience of the utter devastation of so many of the few remaining woods, not just here but in all south-east England, compelled me to write that book, in the hope that people would nurture some of the woods back to life and maybe replace mono-silviculture with native trees and shrubs. Luckily, the aftermath of The Storm also had positive sides: here at least one of the 'black and gloomy' forests came down like ninepins and revealed the potential and wonder of true ancient woodland sites.

A few years after The Storm, the Sussex Botanical Recording Society did a survey here during one day in June that produced two hundred and fifty species of wild trees, shrubs and plants (not fungi, mosses or lichens). About the same time, I heard the irrepressible David Bellamy talking with his infectious enthusiasm about a nature reserve somewhere in the north of England where they had found ninety-three species of plants, which he found exciting. This made me even more determined to produce a book on these woods. If I already had 250 species, how many more were present, and what other biodiversity was here? I could barely wait to finish the other book and start on this project.

Biodiversity

The preservation of biodiversity (much of which has already been lost before we have even become aware of what still remains) is now of crucial concern, not just here but throughout the world. From the outset, my intention was not only to demonstrate how much biodiversity could exist in a relatively small area, but to spur on others to preserve and increase it by trying to reveal some of the wonders our native species have to offer – wonders that may not be immediately or obviously apparent.

We live in an 'instant' society expecting quick fixes and answers for everything. We expect modern science and medicine to instantly cure our ills with miracle technology and miracle drugs. Much of our food is instant and pre-prepared – from burgers, chips, dressings and sauces to whole meals neatly laid out on a platter, ready to eat. Our senses are besieged with instant gratification: our eyes are feasted with ever larger and ever more fantastic imagery in films, on television and on our computers; our ears are bombarded with psychedelic sounds and music that seem to have no bounds.

The picture opposite was taken in Pope's Wood. The four large limbs all belong to a single ash whose trunk is over 3m in circumference. The darker leaves are of the smaller alder trees.

House and southern woods

This is a view, facing south, of the house, gardens and woods beyond. The fields in the distance mark the boundary. The scale in the picture is deceptive: the woods in between are parts of Pope's Wood (page 26) and Butler's Wood (page 40), and the area is much bigger than it looks. The decline into Pope's Wood is quite steep, as is the incline of Butler's Wood. On the right you can see part of Christmas Wood (see page 45).

Nearer to the subject of this book, we seem to hanker after bigger and more spectacular blooms of every hue that will instantly bedazzle our gardens. Horticulturists are not slow in providing them – every year a host of new species and cultivars from every corner of the world emerge and are displayed at increasing numbers of garden shows and shops. We expect instant gardens or instant transformation of existing ones. Modern methods can achieve this. The media does not help: it publishes endless articles and produces television programmes that promulgate just that. I hasten to add that there are also brilliant wildlife magazines and programmes that ought to make us look in the opposite direction.

In the natural world, nothing is either instant or immediately apparent. Trees loom all around you, but how often do you see their flowers? We even talk about flowering trees, referring to the likes of cultivated Wild Cherry, Almond and others all smothered in spectacular blooms, as if the Penduncculate Oak and Ash were in another category. However, they all flower. Maybe they are not as eye-catching, but they are intriguing and beautiful nonetheless. The ground is covered in plants, often so tiny that one either treads on them or walks straight past. There is the minute Common Eyebright, each flower only 2mm or so across and barely visible unless growing in a large clump. The alluring faces of individual orchid blooms are well documented, but have you ever looked closely at some of our more common native flowers, such as the Red Dead-nettle, Bugle or Ground Ivy? They display another world.

Butterflies and moths are usually intent on either disguising themselves or scaring off potential predators. Wild plants do just the opposite – they lure insects by every conceivable means to ensure pollination and their survival. Some have enticing scents and odours, others present bright splashes of colour, however small, that will attract someone somehow, and others are positively devious and murderous, trapping insects sometimes to their deaths in order to achieve their goal (see illustration on page 83 of Lords and Ladies). The Common Nettle makes quite sure that neither you nor most animals will pick or eat it before it has flowered and seeded, by giving a nasty sting or rash.

Of course, I can barely scratch the surface of what happens in this small area, but I have tried, where possible, to show some feature of our trees and plants that you might normally not look for, or to merely display their sheer beauty to tempt you to cherish and conserve them. I am a great believer that pictures can often convey much more than words (maybe I have succumbed to the instant visual fix?), so if this ends up looking like a coffee table book, so be it, as long as even a handful of people will now look twice at our native species instead of dismissing them as mere weeds.

Why native species?

Exactly what is or is not a native species is open to some speculation and discussion. Generally speaking, the wildlife that was present when what is now the British Isles broke away from the mainland of Europe about 6,000 years ago is considered native. Carbon dating and other modern scientific methods have made it possible to identify them, but new facts are emerging all the time. I have used the Botanical Society of Great Britain's list of (native) vascular plants as my criteria.

It seems logical, and nearly always a scientific fact, that those plants and wildlife that evolved together over so many centuries are the ones best suited to support each other in numbers and therefore produce the greater biodiversity. The Pendunculate Oak supports over four hundred species of birds and insects as well as many fungi, mosses and lichens, but the imported Holm Oak supports very few. Our native Yew provides food and habitat for many of our birds whereas the imported conifers provide relatively little.

I have often been asked why I am so keen and even adamant that we should plant and encourage the spread of native species at the expense of imported ones, especially as our flora and fauna are so poor in diversity. I hope I have redressed the first and most important part of this question. The second part is just not true. In fact, the British Isles have more vascular plants per square kilometre than the United States! However, I find such comparisons irksome and meaningless. Just how meaningless is demonstrated by Costa Rica, not much larger than Wales. It has an infinitely greater variety of species (many not yet recorded), than almost anywhere else in the world and scientists flock to study and admire its biodiversity.

There is another side to the question of native species versus imported ones that is potentially threatening and hazardous. If you take a plant (or for that matter any animal or insect) out of its natural environment of climate and soil, some may die, some may become a useful and beautiful addition to our gardens and countryside, but a small percentage will become rampant and invasive, often all but eradicating local species. The damage they do is out of all proportion to their numbers.

It is well known and documented that many of our few remaining natural woodlands, waterways, ponds and other habitats have and are suffering from such invasions. The spread of Japanese Knotweed (*Fallopia japonica*), Rhododendron are just some examples. These plant species were brought into the UK by botanists and latched on to by the Victorians who unwittingly planted them (and some people still do) all over the place, unaware of the effect they would have in years to come. More recently, the New Zealand Pigmyweed or Australian Swamp Stonecrop (*Crassula helmsii*) became widely available in garden centres. This plant is spreading into our ponds where it is taking over to the detriment of all else. Some stonecrop even managed to get into my small vegetable garden pond! They are particularly difficult to eradicate, growing not only in water but also terrestrially.

Alien plants and animals are not the only threat. Species of foreign insects and beetles are potentially even more devastating. They arrived here quite recently through the increasing trade with Asia and other continents. The Asian Longhorn Beetle (*Anoplophera glabripennis*) and the larvae of the Asian Gypsy Moth (*Lymantria disper*) could ravage many of our woodland trees. Everyone knows of the effect of Dutch Elm Disease, a fungus often transported by an alien beetle (*Scolytus multistriatus*). This only affected one tree species – imagine the destruction if these insects really established a grip. One can only hope import controls in this area will be tightened even more.

The reverse is, of course, also true: for instance, I have seen our native Gorse taking over areas in New Zealand and Costa Rica yet here it is not really a problem. Japanese Knotweed is probably not a problem in Japan but it certainly is a problem here. I have also seen it in far-flung Novia Scotia, choking ditches, rivers and waterways at the expense of all else. I hear there are plans to import a fungus that destroys this knotweed, but are we certain that this fungus might not create even worse problems?

Opposite
Shaw Wood and pond in winter mist

Biodiversity the world over is under threat, often for reasons of poverty, so surely it is time we first learn more about our own natural environment before introducing so many new species just to embellish our gardens? Accelerating climate change, or global warming, is bound to bring changes (indeed it already has), but apart from taking some obvious steps, I believe we should allow nature to cope.

I have examples of alien invasions even in my small woodland about which I will write in the chapter The Woods Today (page 20).

I suppose the debate about the Sycamore, which many now consider a native tree (introduced only a few hundred years ago) will rage for decades. I, for one, find it seeds and spreads far too readily and will soon oust smaller native species as well as the understorey. I have a few young sycamores in the woods (inherited from a neighbouring farm), but they are closely watched. If they or their offspring look like spreading far and wide, they will be in for the chop. One pro-sycamore argument is that it has a vast biomass and therefore supports much wildlife (it is certainly prone to excessive numbers of aphids), but the daintier and much smaller native Silver Birch supports many, many more.

When I started this book, I was under the naïve impression that fungi, mosses and lichens were immune to this sort of invasion. Malcolm McFarlane, a moss expert, had only been here a few minutes when he pointed out to me Cape Thread-moss, a moss from South Africa that is showing signs of dominating some of our native species!

Planting and self-generation

Once a diverse wood is well established, it should be left to self-generate. Self-generated seedlings have the advantage that they have germinated in their own time and place and as a result seldom need protection from hungry predators such as deer and rodents. If you *plant* a young tree or flower, the chances are that it will have to be well protected, maybe for several years, until it can take care of itself. Self-sown plants seem to be less susceptible to diseases and predators. I can find no information why this is so, nor, as far as I know, have any scientific trials been carried out. My experience in these woods however, shows it is an indisputable fact.

Creating new woods does present some dilemmas. Does one artificially create biodiversity by deliberately introducing and planting species, or should nature take its own course with no intervention? Ideally, the latter is preferable, but real biodiversity is unlikely to come of its own accord unless on an ancient woodland site. As these woods show, nature needs space and time to truly flourish. But we have no space, so one should intervene if only to prevent one species taking over at the expense of all others. Here on these islands there is virtually no remaining land in which true wilderness can burgeon. My own wood is totally surrounded by farm fields – another situation that must be redressed.

Then there is time. I imagine the debate about self-generation versus planting and human intervention will go on for a long time. But do we allow species to go extinct whilst we argue the pros and cons? Do we allow invasive species, both native and imported, to take over at the expense of all others, or do we intervene?

Managing a woodland for biodiversity

It is easy to see how individuals or organisations intent on the conservation of a particular species could come into conflict with others having different aims. This was true in the recent past, but today a holistic approach is being aimed at. The recent Biodiversity Action Plan has lofty aims indeed, and I only hope that at least some of them can be achieved. But even here I detect omissions. It shows quite clearly with many charts the abysmal decline of all wildlife during the second half of the twentieth century. What it does not show with any charts or even express in so many words is the huge increase in roads, cars, factories, houses, chemicals, and the general increase in wealth, as well as the proliferation of worldwide corporations with immeasurable power. Had it done so, the rise of the latter would have corresponded almost exactly to the decline in biodiversity. Perhaps this approach is too simplistic and direct, but it is nevertheless a fact which society will have to face sooner rather than later if biodiversity is not to become a thing of the past and wildlife an academic curiosity. Vested interests conflict with this Action Plan.

I am concerned with this woodland and not politics, but is there not a parallel here? It is a small, isolated area of ancient woodland, most of which was felled and on the brink of destruction. It could now be a resort, 'chemical' farm or some other modern amenity. It is not protected in any way.

My approach to biodiversity here is very simple: as far as possible allow nature to take its own course. If the soil is healthy, everything else should flourish. Soil organisms and creatures, beetles and insects will proliferate. This in turn ensures a healthy population of plants, birds and other wildlife. In practice and in such a confined area, it is not all *laisser-faire*, but entails quite a lot of work. I have never used any chemicals in the past forty years – a fact which I am sure has contributed to the biodiversity of these woodlands. The garden mostly looks untidy and weed ridden, but the vegetable plot often produces more food than we can handle.

Really invasive species, such as rhododendron, laurel and bracken, and, to a lesser extent bramble and Silver Birch, have to be controlled. Initially, using non-chemical methods may be more work, but I doubt whether this is so in the long run. In any case, the rewards are often immediate and far more satisfying. I have noticed recently that Japanese Knotweed has been chemically treated along the roadsides in the neighbourhood (we have none here, thank goodness), but for at least three years the plant has stubbornly reappeared. The damage the chemicals have done to other plants is unknown. I will show how we deal with similar invasions within the individual woods.

In some ways The Storm was a godsend. It destroyed the Douglas Fir in Owl's Wood in one fell swoop. Unfortunately, it did not also bring down all the conifers in Badgers' Wood, Christmas Wood, nor Morgan's Strip, so reducing their numbers is an ongoing chore.

Conifers are the worst trees for holding back any diverse understorey – not least because their needles cast shade all the year, and fallen needles tend to make the soil inhospitable for other plants. It is no accident, I think, that the UK has only three native conifers, which never naturally appear together in vast numbers. In any case, Scots Pine, although native, is a tree of the North of England and Scotland. We also have to contend with cloned poplars which, like the other invasive species, were deliberately planted.

Apart from redressing the balance between conifers and broadleaves and not using chemicals, I also try to observe a number of other rules:

• No stumps were removed after The Storm, nor are they now unless they interfere with the rides. Many have now turned into amazingly turreted creations (becoming popular in gardens, which is a pity since they will inevitably be removed from woodlands) that have become a haven for mosses, lichens and liverworts.

• Standing or fallen dead trees are left for nature to take its course. Some dead trees have stood for fifteen years – heaven for insects, woodpeckers and maybe bats. Dead trees also inevitably produce a spurt of fungi.

• No rides are cut until late September or October, and the sides of many are left for two years, allowing biennials to flourish.

• Up to now, we have had little problem in keeping the rides wide and light enough for sun-loving plants, but recent wet years have produced an explosion of growth, so many rides will have to be widened. Here we may face some conflicts: do we cut sturdy, good-looking young trees, especially ones in the minority? A possible answer is to skirt round such trees, making half circle glades instead. These in turn should produce tiny microclimates.

• I have not allowed any bonfires on the property since the clear up after The Storm. I soon realised the damage these were causing both above and below ground. I wish I had had the foresight to do what we eventually did with the cordwood in Badgers' Wood (see page 46). We could have mulched it or just left it in piles to decompose, but I doubt whether any clearing and planting grants from the Forestry Commission at that time would have allowed us to take this more time-consuming course! Apart from this, bonfires can be both dangerous and very polluting.

• I fairly regularly have brambles and bracken cleared by either strimming or laboriously pulling them up by hand. Strimming brambles is not good (removal by hand is much better) since emerging tree seedlings will probably also be cut. In any case, it leaves rooted stolons in the ground, so the following year you have four or five brambles instead of one! My obsession with clearing brambles is motivated by the desire to allow smaller, weaker plants to grow and flower (wood anemones and even bluebells will struggle under too many brambles, and others will not flower at all) but this has evoked the comment that I am taking away food and protection for birds. I do not think this is so. Brambles do not flower or fruit on a woodland floor, nor are they thick enough for nesting. In some open areas, we leave them to form dense thickets that provide both food and nesting sites.

• From time to time, I attempt to introduce new species – partly as an experiment and partly to increase the biodiversity, sometimes with strange results (see pages 81 and 91). If I want to augment the existing trees and shrubs, I usually try to consider the birds, who need not only nesting sites but sufficient sources of food. The fruit bearing trees and shrubs are best, and I would like to introduce the Wild Service Tree (*Sorbus domestica*) which bears small, russet apple-like fruits. I also sometimes plant self-sown cotoneaster seedlings (not Wall Cotoneaster *Cotoneaster horizontalis*) from the garden along the sides of rides. This is purely for the birds who really relish the berries, and bees who seem inordinately attracted to the flowers.

About the book

The structure of the book is very simple. After a short history of the land (as much as I have been able to unearth), I have tried to describe the woods today. They divide themselves into eleven very different woods. They vary in their age, habitat or genesis.

Originally I was only going to produce a survey of all the plants, fungi, mosses and lichens, but I soon realised that this would not have shown the true biodiversity, so mammals, birds, reptiles and amphibians, moths and butterflies were added. I toyed with the idea of including just a hint of the beetles and insects, but abandoned it entirely when I read these words by Dr. Jennifer Owen: 'My medium-sized Leicester garden is perhaps more densely planted than many, but in all other respects it is an ordinary domestic garden. Ninety-four species of hover-fly were trapped during the thirty-year period.' This made my mind boggle! In any case, mammals and birds at the top of the food chain would not be present in such numbers if the insect and beetle world were not also very diverse.

Each chapter has the survey, a short text and some illustrations whose sole intentions are to entrap you into conserving our biodiversity. Most of the surveys will be far from complete – that might take a lifetime – but I hope they will give you an inkling of what is not just here but potentially all around you, and all worth preserving. The tree and shrub survey is probably more or less complete, but even here there is no certainty.

All the flora surveys give the scientific name first in alphabetical order, followed by the common name. However, in the fauna surveys I have reversed the order. This is inconsistent maybe, but I feel readers are much more familiar with the English names of birds, butterflies, moths and others. In any case, not all flora have acquired English names and those that have often possess half a dozen or more names depending on which part of the country you live in.

I have probably laid myself open to criticism by including the plants growing in a cornfield meadow that we created on part of the lawn. What has that to do with woodlands? But this book is about biodiversity, and some of these plants, once a common sight for centuries, have become relatively rare. As Charles Hipkin writes in *British Wildlife*, 'Many of our cornfield weeds are former neophytes that have now acquired a positive conservation status'.

My own approach is purely pragmatic and non-academic, but none of the book would have been possible without the many academic experts who have been instrumental in producing the surveys. I have been overwhelmed by their generosity and kindness in giving so much of their time freely and unstintingly with only the environment at heart. The societies they belong to appear at the beginning of the book and the individual names on the acknowledgements page.

Finally, it will have become clear from the organisations who have taken part that the woods are in Sussex, but for obvious reasons I have not given their real names or location. Instead I have made up fictitious ones, some with historical associations. If anyone would like to visit the woods, please contact the Publishers.

Author's note

There is current controversy between various natural history organisations, and their editors and publishers, over the use of capitals or lower case letters when mentioning the English names of flora or fauna within text. At one time, the common names of species always began with capital letters – for example. 'Clustered Bellflower' or 'Scots Pine' – but nowadays many organisations use the lower case for all common names, except for nouns such as 'Scots' which retains its capital. Establishments such as the RHS, RBG Kew, and 'Plantlife', now use lower case for all common names, while other authoritative organisations such as the British Mycology Society, English Nature, and others, retain the use of capitals. To make matters worse, some erudite reference works put all parts of the English name (usually consisting of two or three words) in capitals, and others only the first word! In this book, I have retained capital letters for the English names of species throughout. I have tried to be consistent, but there will still be anomalies. Is it that important when the aim is to conserve and augment our biodiversity?

History

The East Sussex Records Office was my source for tracing the history of the land during the past few centuries. I had already earmarked the areas that I, and others, considered to have been ancient woodland. The generally accepted definition of ancient woodland is that it should have been under tree cultivation, not necessarily native or natural, for at least four hundred years.

I was therefore completely taken aback when I traced the earliest maps that included my forty acres to find that apparently there was not a tree in sight! These maps were published 1724, 1795, 1813 and 1825, and I could find no earlier large-scale maps. Surveys performed for the 1724 and 1795 maps seemed to show nothing but flat fields. Eighteen years later, topography was included, but still no trace of trees. Further research revealed my ignorance of early map making. Apparently early cartographers were usually employed by landowners to show mainly houses and fields that attracted tithe money. Woodlands, especially small ones, rarely attracted any tithe, so they were often not marked as such.

The first detailed map I could find was the Tithe Map of 1843 (the basis for all future Ordnance Survey Maps), and this showed quite clearly that more than half the forty acres had been woodland all along, and substantially bore out what I had already defined as the boundaries between ancient woodland and arable land.

Ancient woodland can be identified by the presence of certain indicator plants, which are marked in the surveys of trees and plants. It was immensely gratifying to find that we had a very large number of these indicators. It is also remarkable that, although virtually all the land is now planted with trees, one can still see a knife's edge between ancient woodland and former agricultural land. The map of 1843 also showed that none of the woodland attracted any tithe, nor indeed some of the small fields.

I confess I found it curious that, during the rich Victorian and earlier land owning times, this relatively tiny plot should have been divided among at least four owners. By the mid-twentieth century, the man who built the present house had acquired more than half the land. The house stands on part of what was a five acre field, for which three shillings (about 15p) annual tithe was payable. A neighbouring farmer owned the remainder. The whole of the ancient woodland was and is named on all Ordnance Survey maps. There were two further owners of the house before I acquired it over twenty years ago. However, it was not until after The Storm that I managed to purchase what we have now called Owl's Wood, Badgers' Wood, part of Pope's Wood and Morgan's Strip, thus bringing this small remnant of ancient woodland under one ownership for the first time in centuries.

Detail of a map from 1795 showing the area of today's woodland that was apparently then all arable land.

Detail of a map from 1813 showing the same area of woodland with only the topography marked.

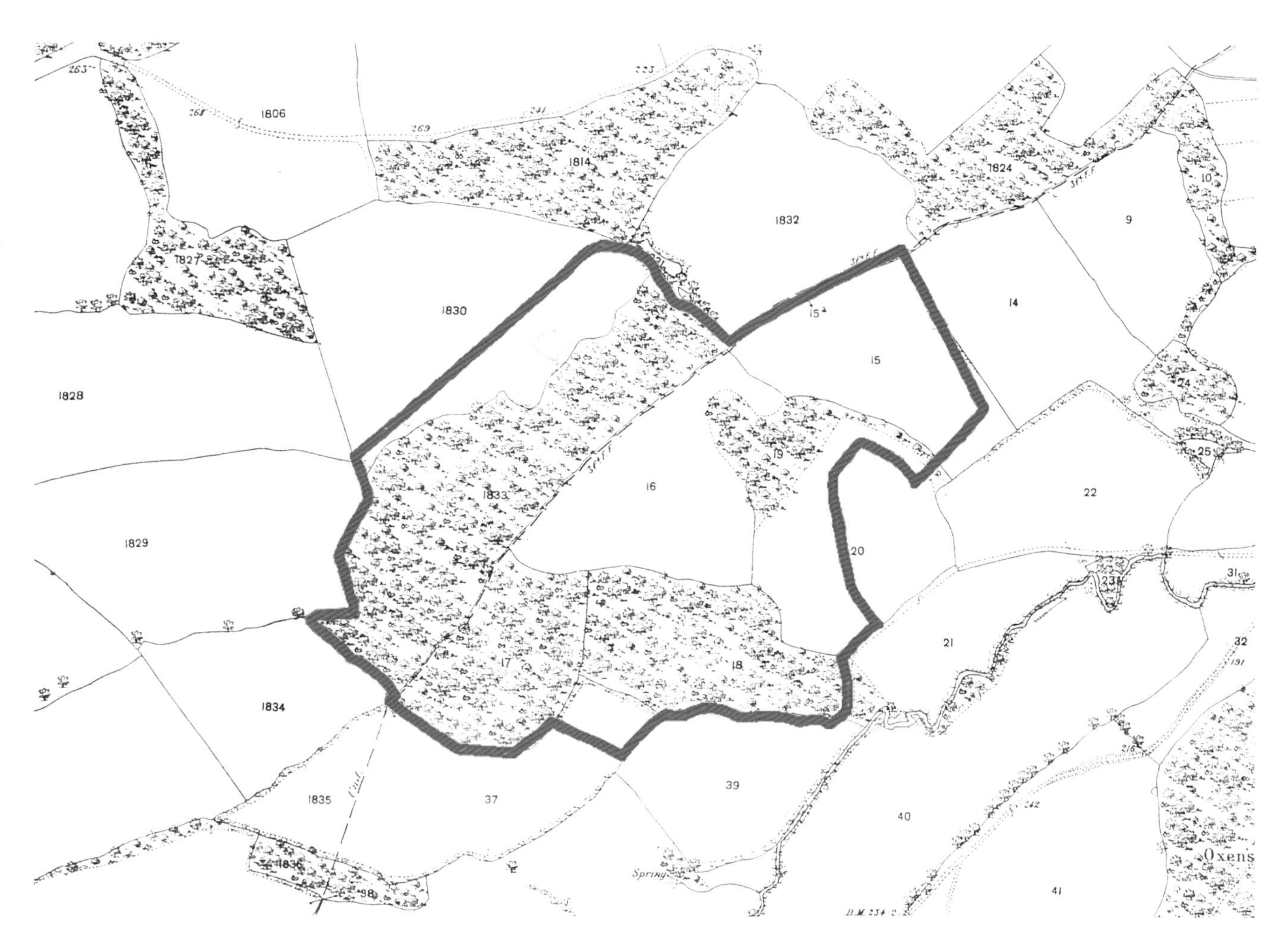

This is our forty acres (outlined in red) taken from the Tithe Map of 1843. You can see quite clearly how, even then, the area was totally surrounded by arable fields. You can also see the area which is ancient woodland, to which has now been added New Wood and Streake's Wood as well as Morgan's Strip. The last three are all arable land on this map. There is no sign of the large pond in Shaw Wood (it is shown on modern Ordnance Survey maps), nor of the railway line which was completed around 1880 and went through the Barn Area. The whole of this single-track railway was dismantled in the mid 1960s.

This aerial view of all the woods was taken at the end of the twentieth century and shows clearly all the natural woodlands that have been added to the ancient woodland shown in the Tithe Map above.
On the right, the railway cutting and large barn can be seen clearly as well as the natural strips of woodland that have grown on either side. These strips are those that have grown over the former railway line, which also has numerous tunnels and bridges. There are three tunnels either on or just outside the small area shown in this aerial photograph. The triangle of woodland on the north is also ancient and is now owned by a farmer growing organic food. So maybe its future is secured.

Railway tracks

This aerial picture shows very clearly one of the railway tracks in the cutting. It is in this area that a number of plants grow that cannot be found anywhere else in the forty acres. This is probably due to the ballast used for laying the tracks.

The Terrain

Although most of the land is now planted with native trees, the forty acres divide themselves into a number of woods, each with a different character, at a different stage of growth and past history. For ease of reference I have given them all names.

All the land is undulating, some quite steep, giving rise to varied terrains and some mini-climates. The lowest point is the valley in Pope's Wood, which is part bog. All the streams and rain drainage end up here. There appear to be several natural springs in the forty acres, one of which supplies us with unadulterated water. It lies on the edge of Pope's Wood.

The soil is all heavy clay on sandstone. In some areas the topsoil is extremely shallow as in Owl's Wood and Christmas Wood, while in others, such as Pope's, there is a deep, lush layer of soil and organic matter. In Owl's Wood one can hardly dig a spade depth without encountering sandstone. It is also the flattest and driest of all the woods.

The main reason for purchasing the house (which is really far too big except when it fills with people and children at weekends) was that its surroundings fulfilled all the criteria for which I had been searching. Not only was it totally surrounded by trees, but also it was isolated and unbelievably peaceful. The house was completed in 1953, and whereas the first two owners had obviously been avid gardeners and 'land entrepreneurs', the third owner had done nothing to either the gardens (which had once been open to the public) or the woods during their eight or so years of occupancy. It is amazing what nature will do in a very short time! I inherited a complete wilderness: the ample vegetable garden (now all under no-dig bed cultivation) was totally overrun with Common Couch (*Elytriga repens*), various bindweeds and brambles, some three or four metres high! The garden fared slightly better since it had been planted with vast numbers of conifers and Rhododendron, which by then had almost taken over.

I loved the woods for their wildness and apparent abundance of wildlife. At that time I was not quite as conscious of the environmental impact of mono-silviculture (the whole of Badgers', Owl's and Christmas Woods and Morgan's Strip were solid conifers) although I cursed its boring and barren outlook. Not even brambles were able to make any headway under dense Douglas Fir in Owl's Wood, whereas in Badgers' Wood some were almost as tall as the Scots Pine! However, during the first few years I concentrated on natural/organic gardening for vegetables, and ended up with twenty-two wonderful raised no-dig beds, which turned the erstwhile clay into beautiful friable, water-retentive and fertile soil. But that is another story!

As I pieced together the history of the land, one sad and disheartening fact emerged: my small forest was an exact mini mirror image of what has occurred throughout the United Kingdom and, indeed, throughout much of the rest of the world. Only tiny pockets of the original ancient woodlands survived. As in my case, most had been felled and replanted with cash crops that gave little heed to wildlife and the dormant biodiversity. In the whole of the UK ancient and natural woodland only account for a little more than one per cent of all the forests and woods. Although over half my small domain is ancient woodland ground, I doubt whether even one per cent exists that has not had major, or at least some, interference one way or another.

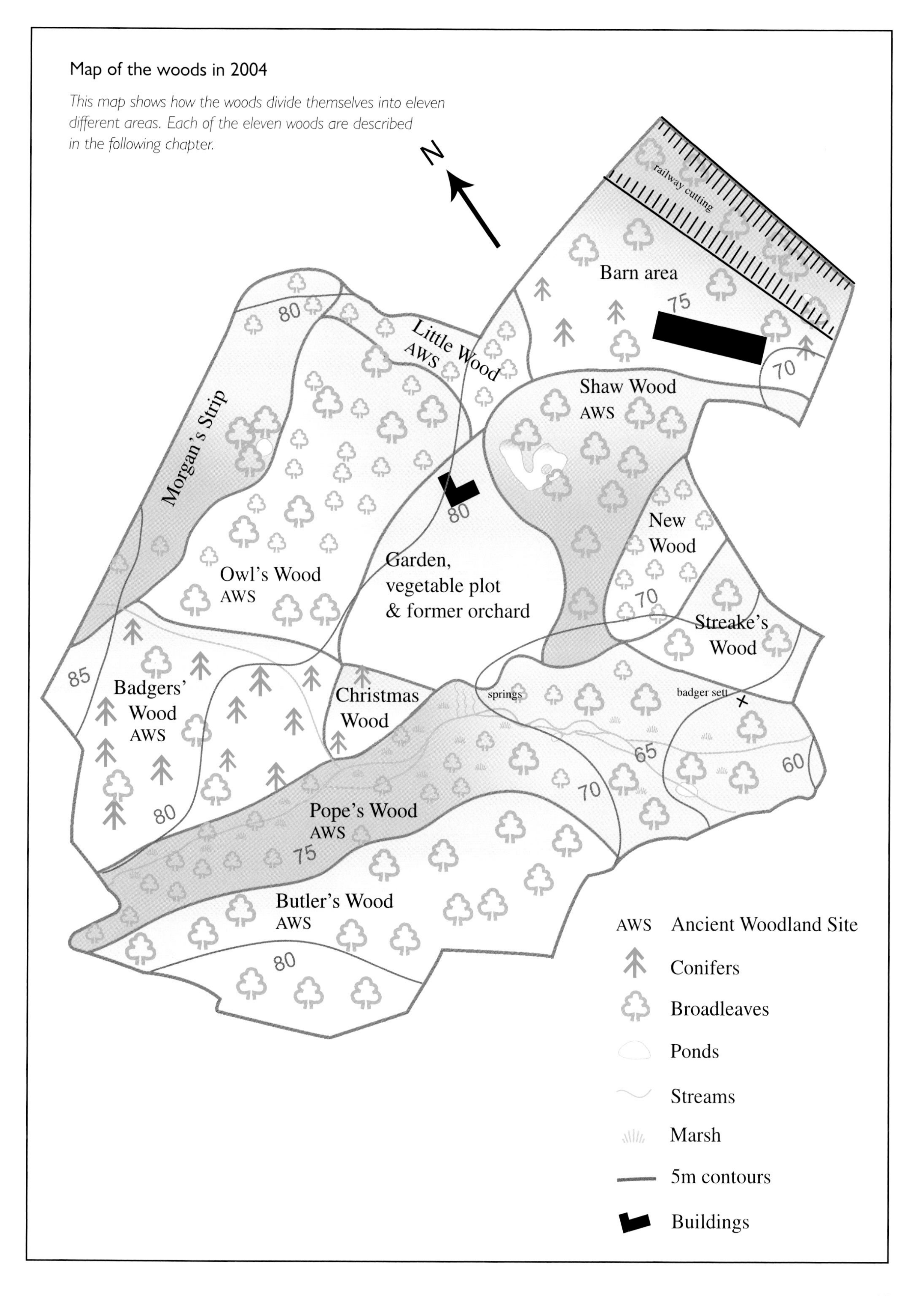
Map of the woods in 2004
This map shows how the woods divide themselves into eleven different areas. Each of the eleven woods are described in the following chapter.
N
railway cutting
Barn area
75
70
Little Wood
AWS
80
Morgan's Strip
Shaw Wood
AWS
Owl's Wood
AWS
New
Wood
Garden,
vegetable plot
& former orchard
70
Streake's
Wood
85
Badgers'
Wood
AWS
Christmas
Wood
springs
badger sett
65
60
80
70
Pope's Wood
AWS
75
Butler's Wood
AWS
80
AWS Ancient Woodland Site
Conifers
Broadleaves
Ponds
Streams
Marsh
5m contours
Buildings

The Woods Today

Shaw Wood

A self-generating ancient woodland with a large pond. Approximately 2 acres.

This is a small wood near the house. Despite the large dominating man-made pond which was constructed between 1935 and 1955, it has many old trees consisting mainly of Pendunculate Oak, Beech, Hornbeam and Field Maple. Unfortunately, several of the oldest trees on the south side succumbed to The Storm. There are many ancient woodland ground plants, including Wood Anemone, Goldenrod, Bluebell, and ever-increasing colonies of Common Cow-wheat. There are also numerous grasses, wood-rushes, fungi and others.

I have a theory that the man-made pond was once the site of an iron mine that was then abandoned. The pond is almost U-shaped and quite large. The channel right around the middle drops suddenly to a maximum depth of three metres or more, whilst the water by the banks is less than a metre deep. I can find nothing to prove this theory. However, there is much iron in this part of the country, and old maps do show a furnace close by and a nearby track that has become the drive to the house. An iron mine could also account for the evidence of coppiced Hornbeam and Alder both here and in Pope's Wood. The irregular patches of Wood Anemone and Bluebell (in contrast to the other ancient woods where there is blanket cover) rather bears out this theory – or at the very least that a great upheaval took place sometime in the past.

Shaw Wood pond in the autumn

A view of the western arm of the Shaw Wood pond in autumn. The whole of the right bank was once covered in rhododendrons, all of which have been removed.

Cow-wheat colony

*This shows one of the large colonies of Common Cow-wheat (*Melampyrum pratense*) that increase annually in this wood. Those on the far side of the path grew within one year of the large Laurel (*Prunus laurocerasus*) being removed.*

The pond also has a man-made island, which is a haven for nesting wildfowl. Unfortunately, crows can still get at the little chicks on the water, but at least the nesting birds are protected from all but marauding Mink. The Coypu (*Myocaster coypus*) has been successfully eliminated from the Broads in East Anglia, so why does no one take similar steps with the Mink? It is not a native species (but has been let loose by misguided animal-lovers), has no predators, can hunt on land, in water and up trees, and does incalculable harm to almost all our wildlife.

I am only the fourth owner of the house. One or both of the first two seem to have had an obsession with Rhododendron and Laurel since they were planted solidly along the track and the east-facing bank of the pond. In 2001 we removed all but a few rhododendrons overhanging the water (shelter for wildfowl). They will have to be cut annually or dealt with in the same way as in Pope's Wood (see page 39). At least now they are not expanding by layering and self-seeding. Common belief seems to be that the soil beneath these shrubs becomes very acidic and poisoned and therefore sterile for several years. This does not appear to be the case here: the charming Cow-wheat spread over this ground in one year, and (rare in these woods) one Broad-leaved Helleborine exists where there were dense rhododendrons. In 2003 the first Common Spotted Orchid appeared where there was dense Laurel. Rhododendron leaves certainly contain toxic chemicals (phenols and di-terpenes) and Laurel contains prussic acid, which converts into cyanide. This is probably the reason why virtually no animal or insect will have anything to do with either of these aliens and could also be one reason why these plants have become so invasive.

The advantage of Shaw Wood is that it is adjacent to both the house and what is now cultivated garden (if you can call my weed-ridden shrubbery cultivated!) so resident and visiting waterfowl can easily be observed. Some years ago, we allowed part of the lawn adjacent to the wood and where there is also a beautiful weeping Silver Maple (*Acer saccharinum*) to 'go wild'. The reward was sudden and unexpected: a large colony of Common Spotted Orchid appeared, a plant which is also fairly rampant in other parts. Their numbers vary from year to year and also depend on how much squirrels and rabbits have decided to nip off the flowering buds. I like to think that this had once been the edge of the ancient wood and that the seeds had lain dormant for over half a century. Wishful thinking?

The south bank, where there was once a giant Pendunculate Oak (brought down in The Storm) and other trees, was left open for waterfowl and most tree seedlings were removed. The result was again beneficial. Large colonies of wild flowers appeared, including St. John's Wort and trefoils, and with them butterflies – many skippers, blues and the Speckled Wood.

In 2003 I introduced two new species of orchid, Southern Marsh Orchid and Marsh Helleborine. The latter succumbed to a hailstorm in the middle of the summer and it remains to be seen whether it will come up again. The Southern Marsh Orchid did not flower that year.

Orchids

*The 'orchid meadow' just outside Shaw Wood. The orchids appeared immediately after we ceased to mow the grass in this area. On the right is the large Silver Maple (*Acer saccharinum*).*

New Wood

A fairly small area, which is not ancient woodland. Prior to The Storm it was full of mature conifers and broadleaves. Approximately 2 acres.

The little plot marked 'New Wood' on the map on page 19 was part of a much larger area that had been arable land for two hundred years or more. Some time during the 1930s it was planted with conifers, mainly Douglas Fir and Scots Pine. A few broadleaves, including some old Beech and Oak, were also present. It was not difficult to imagine old farmland interspersed with some ancient trees. After The Storm only a solitary Scots Pine was left standing!

In 1989 New Wood was replanted solely with native trees and shrubs – a few Douglas Fir and Scots Pine seedlings appeared and some have been left to grow. Grass and the ubiquitous brambles soon threatened to envelop the young seedlings, so the area has been strimmed at least twice. All the young planted trees, here and elsewhere, had to be protected with plastic tree tubes for quite a number of years. Had we not done so, they probably would all have succumbed to rodents, rabbits and deer. More importantly, the plastic tubes served to provide moisture to the saplings during the several years of drought in the 1990s.

It is now a respectably sized young wood containing over a dozen species of native trees and shrubs, some of which are large enough to produce seeds for regeneration and to offer nest sites for birds. Soon, it should be thinned as most of the soil is now so shaded that virtually only ferns survive. So far, none of the plants from Shaw Wood have spread, not even the shade-loving bluebells. Mosses have appeared in abundance on the old tree trunks, and there are colonies of Town Hall Clock and Opposite-leaved Golden Saxifrage. This is unusual since both are ancient woodland indicators. In time it will be a diverse and exciting wood.

New Wood at twelve years old

A view, looking south, of this very young wood, all of which was planted after The Storm.

Streake's Wood

A totally self-generated wood approximately eighty years old. It adjoins farmland, New Wood and Pope's Wood. Approximately 3 acres.

This was part of the same arable land that is now New Wood, and it seems probable that it had been farmland for several centuries. In this case, however, the land was allowed to go completely wild some seventy to eighty years ago. (According to a neighbour, it was farmland until 1926.) When I came here, it was thick, impenetrable scrub. The question was, should we thin it? My advice was that we should, so we went ahead. It is now a 'grown' wood, but many of the trees are spindly and too tall for their girth and therefore liable to come down in high winds.

Many conservationists and environmentalists would argue that the wood should not have been thinned but allowed to take its own course, as this is the ideal way to create a natural woodland, especially as ancient woodlands are adjacent or nearby (Shaw Wood and Pope's Wood). I have no quarrel with this view – in theory at least it would seem ideal. In practice, however, I question its ability to recreate true biodiversity. At present the wood is dominated by only three native tree species: Ash, Alder and Downy Birch. There is also one solitary Penduculate Oak, two limes and, curiously, two

Autumn view

*Streake's Wood in autumn taken from the neighbouring farm field showing the two Copper Beech (*Fagus sylvatica 'purpurea'*) which do not seem to belong to this eighty-year-old self-generated wood. If you look carefully on the right, one of the only two Himalayan Tree-cotoneaster (*Cotoneaster frigidus*) is just visible. I had lived here nearly twenty years before I discovered it! That does not say much for my powers of observation, but the cotoneaster is growing through a lot of dense vegetation and leaning heavily into the field. The other tree grows in Christmas Wood (see page 45).*

Early Purple Orchids

*The rapidly increasing colony of Early Purple Orchids (*Orchis mascula*) which has thrived since we began to clear the brambles every year.*

Copper Beech whose presence is a mystery. There are also a few Norway Spruce, which must have found their way from Christmas Wood. They were puny, stunted specimens when we thinned, but they have now outstripped many of the other trees.

As far as the woodland floor is concerned, this too displayed a lack of biodiversity. There are some wild currants (also present in Pope's Wood) and some lovely colonies of Town Hall Clock, Opposite-leaved Golden Saxifrage, Wood Sorrel and some ferns, but relatively little else. Bluebells – abundant just a small rides-width away in Pope's Wood – have scarcely shown any signs of spreading; the same applies to the Wood Anemone. There is one patch of Yellow Archangel. The rest is in danger of being enveloped by brambles, which, if not cleared, will certainly halt the spread of existing plants and probably prevent new ones moving in. There are quite a number of lichens on the trees that may not have grown if we had not thinned the wood.

I have been very successful in 'propagating' Early Purple Orchids. A few were present on the edge of Pope's Wood and in the middle of the separating ride. Since we annually cleared some of the brambles in the vicinity, they have now spread in abundance, and the more I clear, the more they respond to self-propagation. I am sure many other appropriate species could be introduced (including other orchids).

Only recently have I noticed the emergence of Hawthorn, Hazel and others, and in 2003 I noticed that a solitary clump of Wood Anemone appeared almost on the borders of the farmland. This shows that the Wood Anemone does sometimes, even if rarely, seed itself. Maybe it is also time that we coppiced some of the over-thin trees which in any case tend to come down during storms. This would help the growth of new and existing species.

One thing this wood appears to demonstrate is that proximity plays a far greater part in regeneration of woodland than either birds or animals (see also the section on Morgan's Strip on pages 50-51). However, Streake's Wood has one great disadvantage: the nearest ancient woodland (Pope's Wood) is of a totally different kind. Much of the diversity of Pope's Wood, which is marshy, is not spreading into Streake's, which has reasonably well-drained soil.

I doubt whether Streake's would be the wood that it is today without some human intervention.

Pope's Wood

Ancient woodland, much of which is marsh and bog with small streams. Approximately 6 acres.

This is the kind of wood that is the source of legend and folklore and where imagination can run wild: the stuff of sprites, gnomes and warlocks, spectral magic and mystery. Broken, hollow tree trunks, home to fairies, eye you warily. Tangled undergrowth entwines you, intent on drawing you into the swamp. Mosses hang everywhere and make tree to tree carpets, and looming sedges offer secret stepping-stones through the bog. Flowers abound, including unnaturally gigantic Foxgloves, and glades of shimmering white Hemlock Water-dropwort wait to enfold you in a poisonous embrace.

Biologically this is the most interesting of all the woods not only for its immense plant diversity, but also because fairly large areas seem to have remained undisturbed for almost a century. Even then, the disturbance was only coppicing. Alder was much prized by metal workers, and the furnace not far away may have used the coppiced wood. It would also tie in with my theory of the iron mine in Shaw Wood.

Pope's Wood also suffered relatively little from The Storm. It cloaks a valley into which rainwater drains from both sides, forming a number of streams. The banks on the west and north sides sometimes dry out in drought summers, but the eastern end remains boggy from the overflow of various springs, one of which also supplies the house with water.

The trees in the untouched wet areas are mostly wonderful old clumps of Alder, Downy Birch and willows. Where the ground is less sodden, the trees and shrubs include some huge Ash and Pendunculate Oak. The floor and streams are awash with twelve species of sedge, rush, fern and various mosses, as well as large beds of Lesser Celandine, Opposite-leaved Golden Saxifrage, Bluebell, Wood Anemone, Ragged Robin and many more. There are also several large colonies of Marsh Violet and what amounts to a meadow of Wood Club-rush. In very wet years, masses of Hemlock Water-dropwort appear. There is also a fascinating little pond (reputedly owing its existence to a 'doodlebug' that came down in World War II) teeming with all kinds of wildlife, including newts.

At the far eastern end, on a bank at the edge of Streake's Wood, there is an old badger sett. I call this the old Badger sett, as there is another one, much larger and more recent, in Owl's Wood; see page 48. The old one has been there all the twenty-odd years I have lived here, and probably long before that. It is well placed, facing south and near one of the tiny streams that abound in this wood. Badger trails can be seen throughout Streake's Wood and up into New Wood and Shaw Wood. The badgers themselves can often be seen in the garden and sometimes sneaking around the chicken run in the Barn area. I put down newspaper covered by wood chips in the shrubbery to keep out some weeds (mainly rather obnoxious creeping thistles), but, alas, they do not last long – the worms underneath are too much of a temptation for badgers. Paper and wood chips are flung everywhere. I cannot blame them solely – rabbits often try to dig new burrows there too.

Deer wander through all the land, including the lawn, but their hoof marks can be seen most prominently along the shallow sides of the streams in Pope's Wood. It is also here that we spotted a footprint that could have belonged to a Water Rail. It is the right habitat for this very shy bird but regrettably none of us has ever spotted one.

Unfortunately, not all the wood is like this. The extreme western end was felled and re-planted with poplar hybrids about fifty years ago by the previous owner in the expectation that these poplars (they also appear in parts of Badgers' Wood and Owl's Wood) could be sold for making matches. This never came to fruition – hardly surprising since, according to the company, Swedish Match, matches have not been made from wood grown in the UK since 1983 and the production of matchsticks ceased here in 1994. Thus, their manufacture had ceased by the time the trees were mature. So I inherited these trees, which have grown enormous. They are singularly uninteresting in that, like all cloned commercial timber, they produce straight, tall trunks with open branches that are relatively small. Like our native poplars, they are dioecious (male and female flowers are on separate trees), and the large red male catkins closely resemble those of our indigenous Black Poplar. In all other respects, however, they have neither the character nor the unpredictability of our native Black Poplar, which is now fairly rare and which progresses into a stunning if eccentric tree in maturity.

These 'matchstick' poplars also multiply with ease. This, and the fact that they are uninteresting non-native trees and seem to be unduly draining the soil, made me decide that their days should be numbered. We have, therefore, begun to fell and/or ring-bark them. We shall leave a few good specimens for 'old times sake', but I am afraid the rest are doomed. I have no idea what to do with the wood as it does not seem particularly suitable for furniture (I think they do make veneers from hybrid poplars) but, I am told, it will burn quite well if left for several years. I am sure our hungry wood burner will devour them one way or another! Some of the ring-barked trees will be left standing. Beetles, insects and woodpeckers can have a ball on them. The trees remained green for nearly two years after they were ring-barked and the butts are going to coppice unless something is done about them! The space left by the poplars will largely be left to self-generate, but it seems an ideal area into which a handful of the native Black Poplars should be planted.

The north-facing bank above the hybrid poplars also looks too clean and tidy. It was therefore probably felled along with everything else at the western end of the property. Nevertheless, there is much Hazel, Dwoney Birch and Alder on this bank, and in spring it becomes a riot of bluebells and wood anemones.

In the 1950s, an outdoor cement swimming pool was constructed at the south-eastern end of Pope's Wood. It was well done, with a combination of dams and sluice gates that could constantly supply the pool with fresh spring water from up-stream. Unfortunately, it was not maintained – the cement cracked at one end of the dam, so now there is only a shallow silted pond that, nevertheless, supports wildlife. This includes newts and some interesting dragonflies. Maybe one day, given time and money, it could be restored as a real pond. Also, large numbers of Laurel (some of which have grown into trees) and Rhododendron, as well as conifers, were planted all around the pool. So far the conifers have been left, but we are doing our best to eradicate the Rhododendron.

Whichever owner of the house had an obsession with Rhododendron and, for that matter Laurel, did not confine their planting to the garden, shrubbery or swimming pool area. They also planted some along the main stream in the middle section of Pope's Wood. These, too, were spreading alarmingly so we have cut them and left them in large piles to rot and provide habitats for wildlife. We were faced with having to cut or strim them annually until the Forestry Commission came to my rescue. They are financing the purchase of thick industrial plastic sheeting with which the stumps will be covered and the plastic anchored down. The plants should die within two or three years (they have done so wherever else I have tried this method), when the plastic sheets will be removed. In the meantime other plants are already springing up where previously none grew under the dark shade. Here also, the ground does not appear to have been 'poisoned'.

Experiments are underway in the UK to reintroduce the European Beaver (*Castor fiber*), which was resident in Britain until the sixteenth century when it was hunted to extinction. Progress seems to be slow, but I cannot help wondering what might happen if beavers were let loose in the old swimming pool. They would soon have it dammed! And they would find plenty of trees to their liking: Alder, Oak, Ash, Rowan and others. But I fear Pope's Wood is too small unless they were allowed to expand into neighbouring farmland and the river nearby. It would certainly make ideal beaver habitat!

'Enter these enchanted woods
You who dare...'

George Meredith, *The Woods of Westermain*

Ancient Alders

Huge clusters of many-trunked Alder (Alnus glutinosa) *reign supreme in the middle section of Pope's Wood. They seem to have been coppiced some eighty to a hundred years ago. There are many dead hollow trunks that give this wood its special mystical, eerie qualities. Mosses abound, often growing high into the canopy. It could be argued that we should continue coppicing, but I think the wood may lose much of its ancient mood and quality. In any case, at present it seems to maintain its own balance between shade and light – trees fall or die and are quickly replaced by sedges and Wood Club-rush followed by young trees. This is not unlike in a virgin tropical rainforest where there are always plants just biding their time to shoot up and blossom.*

Rivulets of blood and Ents looming up behind?

One of several shallow streams running through this wood showing the presence of iron.

Greater Tussock Sedge

*One of the glories in Pope's Wood is the amazing Greater Tussock Sedge (*Carex paniculata*). Some of the sedges are very old with great, soft trunks almost a metre tall. They do not have the graceful flowers of the Pendulous Sedge (*Carex pendula*), which is widespread, but possess an antique dignity all their own. There are many other less spectacular but no less interesting sedges and rushes throughout this wood.*

Hemlock Water-dropwort

*In some years leas of Hemlock Water-dropwort (*Oenanthe crocata*) dominate the open areas. Some grow between one and two metres tall. All parts of the plant are poisonous.*

Giant Foxgloves

If you are a Foxglove (Digitalis purpurea) *and want to survive in this wood, you sometimes have to grow as tall as humans!*

Stream and Anemones

Another stream in this wood, this time running clear – almost blue. Note the Lesser Celandine (Ranunculus ficaria) *and Wood Anemone* (Anemone nemorosa) *growing randomly. In drier spots there are also Bluebell* (Hyacinthoides non-scripta). *Compare this picture with the lovely but clinical appearance of the wood anemones in Badgers' Wood (see page 48).*

Marshy scene

*Another typical marshy scene in Pope's Wood. Note the Wood Club-rush (*Scirpus silvaticus*) in the foreground and the abundance of mosses, some even hanging from the tree branches. The Wood Club-rush has become abundant in recent years. It has colonised part of the old swimming pool and in one clearing it has virtually created a 'meadow'.*

'Doodlebug' Pond

The small pond in Pope's Wood teems with plants and wildlife. When this picture was taken, some of the rampant Rhododendron were still present. They may look pretty here, but it would not have been long before they encroached right around and eventually into the pond. They are no more! This pond reputedly came into existence from a doodlebug crater in World War II.

Condemned Rhododendrons

We cut all the Rhododendron (and Laurel), of which there were many in this wood, but within a short time they re-emerged. The stumps have now been covered with thick commercial black plastic (see inset). The plastic has been anchored down with the old branches and leaves. In two or three years all should be dead, never to regrow! Normal thin plastic would not do the job – tiny holes would be inevitable, and most plants only need the tiniest chink of light to force their way through. I have tried plastic mulches from garden centres, but these are no good – in less than a year plants have pushed their way up where the mulch has become degraded.

Opposite

Aerial view of ring-barked poplars

This aerial view of part of Pope's Wood was taken in the autumn of 2003. In the foreground are the cloned poplars that were left standing and their dominance can be seen quite clearly. In the top left, you can see the Scots Pine in Badgers' Wood, and in the bottom right hand corner the edge of the Beech glade in Butler's Wood is just visible. The poplars had been ring-barked for nearly two years but remained green until the drought months in 2003. They now appear to be succumbing, but unfortunately many of them (both the felled and ring-barked ones) are sprouting from the base – something that will have to be dealt with in the future if we are going to avoid having another forest of these uninteresting trees. Most of the ring-barked trees will be left standing – good for fungi, beetles, insects and, of course, woodpeckers in years to come.

Felled Poplars

This is a sight I do not like to see in any wood – large felled trees – but the dominance of these cloned poplars was, I believe, becoming just too much for an ancient woodland site.

Butler's Wood

All ancient woodland, much of which was destroyed by The Storm. Approximately 6 acres.

This area, most of which is a north-facing slope, seems to have been left undisturbed until The Storm, which, to my great regret, completely flattened most of the wood. Only the far western corner and a strip on the south boundary remained. Here, some lovely old native trees still grow. They include mature Wild Cherry and, strangely, some English Elm saplings, which at present are greatly overshadowed by large Oak and Hornbeam.

Many of the stumps that were not totally ripped out of the ground by The Storm have regrown as coppiced trees. These are mostly Hornbeam, some of which have turned into a great sprawling mass of branches. There is one Beech that has grown into a great bush – it looks quite unusual since in the past these trees were usually pollarded, not coppiced. Also, coppiced Beech does not normally regrow.

One small corner, at one time owned by the same farmer who had planted the poplars further down in Pope's Wood, is occupied by a grove of young Beech which must have been planted in the 1960s and also survived The Storm. They are now sizable trees that, luckily, have so far survived the attentions of the grey squirrels. But I wonder why only a single species was planted and why the trees were planted in tedious straight rows? The trees are too few to be of commercial value and are now of a size that inhibits the self-generation of other species. However, each spring the grove and the surrounding woodland burst into a haze of blue.

The rest of Butler's Wood (mainly the bank sloping into Pope's Wood) has been replanted and includes some species that, as far as I am aware, were not previously present in the forty acres. These are Wild Service Tree, Whitebeam and Small-leaved Lime. The first two are doing very well and obviously love the soil and terrain. Even some of the Small-leaved Lime is thriving. A small glade of Sweet Chestnut, not native by the way, was also planted.

I left one replanted strip between two rides on the slope of this wood and did nothing in the way of maintaining it by either thinning the birches or rescuing the

One or six Ash trees?

*The original Ash (*Fraxinus excelsior*) may have died, come down in a storm or even been coppiced too closely. Whatever the cause, there are now six clones.*

small trees from brambles. Virtually none of the planted specimens survived; instead, there is one enormous coppiced Hornbeam straddling the whole strip, and many birches and hazels, neither of which were planted. It is now very dark inside the strip, so there are few brambles but a number of ferns. It is a haven for fungi in the autumn.

In parts of this wood, bracken is proving a problem. It destroyed many of the planted trees and inhibits any natural regeneration. It has to be cut annually, twice when time permits, and is now weakening. By midsummer it is only 10-15 cm tall.

In spring all six acres of Butler's Wood are blanketed with spring flowers – only where the Birch has not yet been thinned are they still absent.

Snow scene

*Snow is now a rarity in these parts but, a few years ago, we had a fall that lasted just long enough for this picture to be taken. It shows the small grove of Sweet Chestnut (*Castanea sativa*) that was planted after The Storm. This is one of the highest points in the wood where they would have reasonable drainage. At the time of the picture the Sweet Chestnut were about ten years old. Initially, some thinning of Silver Birch (*Betula pendula*) had to be made.*

Bluebells in Butler's Wood

*A few areas in Butler's Wood were almost untouched by The Storm. This is part at the top of the wood, adjacent to the Beech plantation (overleaf), and every spring it is smothered in Bluebell (*Hyacinthoides non-scripta*). In time, the slope, which was completely decimated in 1987, will be as thickly populated with spring flowers. At present the birch saplings are too thick even for these tough ancient woodland indicators. Birch has already reappeared in the Sweet Chestnut grove pictured on the previous page. Much of this wood is very well drained and one can only hope that the tough bluebells can withstand very dry summers. During the very hot, dry years of the 1990s virtually no seedlings survived. The section of wood in this picture includes not only grotesque wood sculptures, as seen in the foreground, but also one of our oldest oaks (see photograph page 71). There are also a number of straggly elm saplings indicating that some must have been present many years ago. I am uncertain whether to give them light in which to flourish or wait until all danger of Dutch Elm disease has passed. The beetles that transport the fungal disease (*Scolytus scolytus *and* S. multistriatus*) will only attack fairly robust trees.*

Dead Silver Birch

Wherever possible, we leave dead trees for beetles and woodpeckers. This Silver Birch has been standing for a very long time. (I begin to wonder whether it was really Birch since they normally rot fairly quickly.) Note the bracket fungus which appeared years ago and is still there today. It will be interesting to see what happens to the ring-barked poplar hybrids that will be left standing.

Overleaf

Beech grove

This should really be described as 'the Beech alley' rather than a natural grove. In 1981 the trees were saplings a metre or so tall. In spring the floor is covered in bluebells, but during the rest of the year nothing much grows in the understorey – the Beech canopy is now too dense and the ground is just too shaded. If you ever plant trees, do not plant them in straight rows, and never plant only a single species like here – it looks unnatural and will not produce any kind of biodiversity. Note also how the Beech have not developed many lower branches – a natural Beech in space will develop lower branches that are horizontal or even sweep down to the ground.

Christmas Wood

About 1 acre of solid Norway Spruce.

This little wood is just plain boring and, frankly, an eyesore. Approximately forty years ago a plot of agricultural land was planted with Norway Spruce (*Picea abies*), the traditional Christmas tree. It was never thinned, and when I acquired the wood it was so dense you could not even squeeze through without being whacked by ugly, needleless branches. The ground was nothing but brown needles all the year. Currently only about half of the trees remain, but the canopy has spread so much that even now only brambles and fruitless elderberries grow.

A small area at the edge bordering Owl's Wood was clear-felled. This was immediately colonised by Rosebay Willowherb. A number of broadleaves are now also thriving, and underfoot vast numbers of Bugle, Primrose and other woodland flowers are spreading. These are all plants that do not rely on mycorrhizal fungi for their existence (see page 108). Potentially it could be an attractive wood with a great deal more going for it than at present if all the spruces were removed. We are still slowly thinning them – the split trunks make wonderful retainers for my raised beds in the vegetable garden! But for the time being, total clearance will have to wait until time and finances permit.

The wood has just two redeeming features (apart from the small part that was felled): at the edge, bordering Pope's Wood, there is a gigantic Japanese Larch which probably only survived The Storm because it was protected by the spruces. The other is curious: a cotoneaster, almost as tall as the Norway Spruce, grows near the orchard boundary. I believe this must be the Himalayan Tree-cotoneaster, introduced to this country in the early nineteenth century. It is a bit straggly but improves as trees around it are removed. Its presence is a mystery since there is nothing else like it anywhere in the garden or in the neighbourhood. In fact, I have rarely seen this tree anywhere else. But, having puzzled about it for many years and having just written this, I discovered yet another smaller specimen struggling among the ashes, birches and brambles on the very edge of Streake's Wood (see page 24)! A good lesson in not making any dogmatic statements about the presence or absence of any species!

*There is really nothing much to be said about these Norway Spruce (*Picea abies*). Although we have removed almost half of them, the area is still just brambles and a few elderberries, none of which ever fruit. I hope, some time soon, we will be able to remove all but a few good specimens and allow natural regeneration to take place.*

Badgers' Wood

Ancient woodland totally felled some time in the 1950s and densely planted with Scots Pine. Approximately 5 acres.

This wood, together with Pope's, Butler's, Owl's and Little Wood made up the solid tract of original ancient woodland. It saddens me to see the enormous butts of Pendunculate Oak and Sweet Chestnut that still remain. The 'blanket' of Scots Pine was never thinned! The whole plot slopes gently southwards, but there is also a steep westerly slope. A stream runs down the west side. This is mainly drainage from the surrounding farm fields about which I was not at all happy since the water contained evidence of excessive amounts of chemicals. The farm has recently changed hands, so maybe things will improve. This stream dries out in summer, but there are also a number of wet patches that are, I suspect, further springs.

When I purchased this wood together with a small part of Pope's Wood, the whole of Owl's Wood, Morgan's Strip and Little Wood in 1989, it was totally overgrown with brambles, some as tall as the smaller Scots Pine! As far as I could make out (it was impenetrable) hardly anything but these two species grew there.

Unfortunately, not a single tree came down during The Storm! At the behest of the Forestry Commission (which had financed the clearance and replanting of all the other destroyed woodland) we set about clearing, thinning and creating rides throughout Badgers' Wood. Rather than burn the brambles and cordwood, we made tall long 'snakes' of them all along the rides and had these chipped into mulch by a fantastic machine, a Rousseau Mulcher mounted on a Same tractor. At that time, in the early 1990s, there were relatively few of these machines around. Now there are plenty, and wherever trees have to be cut in numbers they are used to make useful mulches. However, as soon as we humans are on to a good thing, we seem to go to extremes: these machines are now creating mountains of fermenting mulch, with unknown consequences.

In our case, the cordwood and the relatively little Rousseau produced sensational rewards. Wherever we could keep the brambles from becoming too dense, swathes of woodland flowers suddenly appeared, the most widespread being wood anemones and bluebells, but there are also vast numbers of Enchanter's Nightshade, Lesser Spearwort, Foxglove, ferns and many others. In addition, a large variety of native broadleaved trees and shrubs have germinated. A few reasonably sized Pendunculate Oak and Ash and one Willow had survived among the dense Scots Pine, but these have now been joined by Hornbeam, Hawthorn, Hazel, Elderberry, Silver Birch and others.

A wood transformed

This shows part of the wood not long after the Scots Pine had been thinned. The floor underwent a miraculous transformation – from skyscraper brambles to this! Some broadleaves are now making their presence felt. However, many more Scots Pine will have to be felled before this becomes anything close to a natural woodland.

Rather surprisingly after so many years, some of the Sweet Chestnut butts sprouted as soon as they could see daylight, and in some cases they have already become fair-sized trees. Annoyingly, some of the 'matchstick' poplars are also present (not a single one came down in The Storm!), but they are doomed.

Although we felled a large portion of the conifers, we are now faced with having to take out even more. The canopy has spread so much that in places it is inhibiting the germination of other trees, shrubs and other plants.

The rewards of thinning and mulching did not stop at new trees and flowers: eventually the thick blanket of wood chips along the rides rotted and produced vast numbers of easily accessible earthworms, so the wood has become the badgers' daily hunting ground – hence the name. In fact, I believe this vast larder was instrumental in the formation of a second sett. It is actually located in Owl's Wood, on the bank of the stream separating the two woods. Initially, I am sure it was a large rabbit warren, but is now definitely inhabited by badgers and is enormous – some thirty metres long.

Opposite

Aerial view

A view of part of Badgers' Wood from a helicopter. It shows clearly how many conifers still remain despite the fact that at least fifty per cent were removed. The new young broadleaved saplings are not, of course, visible – only those that managed to survive the conifer planting fifty or more years ago. The group of deciduous trees on the right marks the small stream that divides this from Owl's Wood. Here, probably because it is much wetter, the conifers fared less well, but further down there are still many, including Douglas Fir, that will have to be felled. The group of broadleaves also includes a number of the cloned poplars, which have now been ring-barked. You may be able to identify them by their even, conical shape. This boundary between the two woods is also the site of a very large badger sett.

Regimented trees

Whichever way we thin the Scots Pine, we cannot get rid of the regimented look of a commercial plantation. The reappearance of the Anemones was a joy, but compare this with the wildness illustrated on page 35.

Reborn Sweet Chestnut

I was always saddened to see the many huge butts that remained in this wood, even fifty years after felling. This is one of the Sweet Chestnut trunks that sprang to life after some of the Scots Pine had been thinned. Unfortunately the same did not happen to the many oaks.

Morgan's Strip

A strip of land on the northern boundary that was agricultural for over 200 years until 1952. Approximately 3 acres.

Until the spring of 2003, this was more or less as we found Badgers' Wood, with one major exception: it was never ancient woodland. As far as I can make out, it had been arable land for several hundred years, but just this strip was planted with Scots Pine about the same time as Badgers' Wood. One can trace the boundary between this once arable land and the neighbouring ancient woodland of Owl's Wood quite clearly, since there are the remains of old ditches between the two. Ancient woodland indicators abound in Owl's Wood, whereas in Morgan's Strip there are virtually only brambles and some ferns. In just one or two moist places, bluebells have begun to spread over the boundary, but by only a mere two or three metres.

On the border of Morgan's Strip and Owl's Wood, and therefore at the edge of the former arable land, there is a reasonably sized pond, round which Penduncalate Oak (several very large specimens), some willows and blackthorns grow. It must once have been a dew pond – it is certainly the right shape and size. It is the only pond that appears on the Tithe Map of 1843 (see page 17). It frequently dries out in drought years, and the water looks stagnant, but nonetheless moorhens use it to nest occasionally. Insect life abounds, a fact that is, no doubt, appreciated by the many resident bats – pipistrelles can be seen flying round it at dusk.

In 2003, the Forestry Commission agreed to part finance the felling of the Scots Pine. There was evidence that a large number of native species of trees and shrubs were attempting to grow through the brambles but were held back by the dense canopy of conifers. Here and there, on the north boundary, some sizable oaks, hawthorns, blackthorns and ashes had managed to push their way through. So far, we have found twelve native species of trees and shrubs, the seeds of which have undoubtedly come from Owl's Wood. Unfortunately, I cannot show the results of this clearance before this book is published, but I am certain that in a very short time Morgan's Strip will turn into a diverse natural woodland, unlike Streake's Wood which has taken nearly a century to show any real biodiversity. This will surely be due to the fact that Morgan's Strip is narrow and has a very long border with ancient woodland and the terrain is very similar to Owl's Wood.

Felling

This is the widest area at the north-eastern end of Morgan's Strip. A number of ashes had taken hold where there were virtually no pines. The amount of timber produced from relatively few puny Scots Pine was extraordinary. I imagine most of them will end up in our wood burner, but a number will be turned into wood chips – invaluable in the garden.

Dew pond

These two pictures show the dew pond before and after the removal of the Scots Pine. The difference in the background is amazing; the picture below was taken only a matter of weeks after clearance had taken place. In that short space of time you can clearly see how the emergence of light has turned everything green. Unfortunately, the brambles also have had a new lease of life! The pond surface is already benefiting from the light. The picture below also illustrates the effect of the drought in 2003 – in the picture above, the Pendunculate Oak on the far side of the pond (one of the oldest oaks in the woods) is clearly growing out of the water, whereas below, it is standing high and dry. In fact, it was surprising the pond did not dry out that year – it often does so in dry years.

Owl's Wood

Ancient woodland until the 1950s when it was felled and planted with Douglas Fir. Approximately 5 acres.

This was another unfortunate ancient woodland that was completely felled (about the same time as Badgers' Wood) and planted solely with Douglas Fir. These conifers came down like dominoes in The Storm. Only a handful survived, and one or two are now massive trees. Douglas Firs are one of the most popular commercial trees, not only because they grow astonishingly fast, but also because the wood is good for a variety of uses. It is also harder and tougher than most other conifers. A number of the 'matchstick' poplars managed to stray into this wood and survived The Storm. They will not be there much longer!

The land under Owl's Wood is the flattest of all the forty acres of my land, but still slopes very gently south-eastwards, much of the rainwater draining into the pond in Shaw Wood. The soil is fairly shallow – one only has to dig to a mere spade depth before hitting the sandstone.

Owl's Wood was replanted with only native species. I thought we had introduced Alder Buckthorn only to discover as late as 2003 that some were already growing on the boundary of Pope's Wood and Butler's Wood! Later, I introduced a few Wayfaring Tree (not already present), but they are growing slowly and each year get eaten by minute caterpillars whose identity has eluded

Sudden emergence of Bluebells

For nearly six years I had to view the monotonous,dense Douglas Fir and barren forest floor that were quite close to the house.The devastation of The Storm was as complete as the subsequent transformation. On the right you can see the edge of Little Wood that totally escaped The Storm. The trees in the centre and left are mainly those round the pond in Morgan's Strip that also survived. Many trees (about four thousand) were planted here and elsewhere – a rife scenario for those environmentalists who argue for total self-generation. In reality, this wood is turning out to be 50–75 per cent naturally generated. It seems, to me, that I have managed to get the best of both worlds.

me. The Wayfaring Tree prefers chalk and limestone, of which there is none here. This probably accounts for their slow growth. I also introduced a handful of Dogwood and these also are not exactly thriving, although some planted along the pond in Shaw Wood (to replace the rhododendrons) are faring better, probably due to the extra moisture. If neither species survives, I shall not persist – it is useless trying to push a wild plant if it does not like the terrain or conditions.

Ash trees are self-seeding and survive in vast numbers elsewhere, but the planted ones have not done well in Owl's Wood. This is strange, and I can only put it down to the fact that it is drier here, despite the moisture-retentative clay. Or maybe they just did not like being transplanted. Only a few metres away in Little Wood, I have to keep the Ash seedlings in check. I tried transplanting some into Owl's Wood without giving them any protection. Rabbits ate every single one, yet nearly all those in Little Wood survive!

One of the problems in Owl's Wood, and elsewhere where the ground was cleared after The Storm, is the vast numbers of Silver Birch, which grow so dense that you cannot even put a hand between the seedlings. This is another case where some environmentalists will argue that they are best left alone and nature will sort things out in eighty or one hundred years. This may be true if huge areas are available, but here in Owl's Wood, a mere few acres, I doubt whether it would ever produce a true diversity of species without some management. A few small Pendunculate Oaks managed to survive both the planting of the Douglas Fir and The Storm and with the sudden advent of light soon spread out into sizable trees. Young oaks are germinating everywhere in clear space but not at all where the Silver Birch has been left untouched. Unfortunately, most oaks germinate in the rides, which are cut most years. As far as any of the *planted* trees are concerned, only those that were given a little space and light by clearing have grown sturdily. Where the birches were left untouched, there are now virtually none of the planted species. The converse also seems to be true. At the top of Butler's Wood where there was no slope, a number of the original trees survived – mostly Pendunculate Oak and Hornbeam. These are both producing seedlings, yet I cannot find any Silver Birch.

There is a further point: if there are literally hundreds of Silver Birch growing in a few square metres (we counted between twenty and thirty in just one), surely it is better to thin them to just a handful? These will grow sturdier and faster, and in a mere ten to fifteen years these trees will display all their beautiful traits, whereas if left untouched they will turn into emaciated sticks. I know this may sound a bit cosmetic, but those that remain will more readily survive storm and drought, and other species will grow immeasurably better.

I mentioned in the Introduction that the survey of trees, shrubs and climbers was probably complete, and in the survey I originally indicated that Aspen had been introduced by me. Yet in the summer of 2003, when some of this book had already been written, we discovered four sturdy Aspen whilst thinning some of the Silver Birch in a small area in this wood. The Aspen were not easily visible from the ride but had outstripped every other tree! They were certainly not present after The Storm. I think this picture shows how total was the devastation.

I called this Owl's Wood because, whilst the Douglas Fir were still standing, I spotted on several occasions a Barn Owl, clear and unmistakable against the dark background. I have only seen it once since (I have not actually searched for it at night), but one can sometimes hear its hoarse, eerie, prolonged shrieks.

Foxgloves and wild flowers

These two pictures show more of the post-Storm transformation that took place. (It is a pity that I have no picture of the gloomy Douglas Fir forest.) The foxgloves no longer have such a wide open expanse in which to colonise. The Yellow Archangel, Red Campion, Stitchwort and others in the picture on the left tend to grow on the extreme southern boundary of this wood. I believe this could be because the remainder of the wood is too well drained and dry. Also, there is evidence of a spring in this area.

Aspens

For years I was firmly under the impression that we had no naturally growing aspens. It was not until we thinned a part of this wood in 2003 that we discovered four sturdy aspens that had outstripped all the other trees, either planted or self-generated. The picture shows three of them. What a difference between this and the picture on pages 52–53. It also shows the rate of growth from nothing in twelve years.

Little Wood

Self-generated ancient woodland. Approximately 1 acre.

This is a little gem and as interesting as Christmas Wood is boring. It is really part of Owl's Wood, but for some reason, although it was felled at the same time as Owl's Wood (there are still some large butts visible), it was never replanted. It is now a totally self-generated young wood, but unlike Streake's Wood which was once pasture, this was ancient woodland, and the result is a very different and diverse wood. With the odd exception, all the plants are native. The species include: Hawthorn, Hazel, Ash, Oak, Silver and Downy Birch, Beech, Scots Pine, Hornbeam, Field Maple, Wild Cherry, Goat Willow, Alder, Sweet Chestnut and Blackthorn. (Incidentally, Scots Pine is not normally considered a native to Southern England, and Sweet Chesnut only appeared here about 2,000 years ago with the advent of the Romans.) For such a small area it shows remarkable biodiversity. The floor is covered with an impressive array of ancient woodland indicators and other plants. When I purchased the wood, it was really dense and most trees were very spindly. I thinned it about ten years ago. The trees subsequently branched out, and the wood is rapidly turning into a good example of a naturally-generated wood.

A few Douglas Fir seeded themselves and rapidly outstripped the other trees. Only one or two have been left. I also planted one stunning Golden Oak (*Quercus rubra 'aurea'*). It is still very small and grows so slowly that I doubt whether it will be a sizable tree in my lifetime. This species requires full sunlight to grow into all its glory. I also planted a number of wild daffodils around this tree but, although sold as wild daffodils, I doubt whether all are native. As always, brambles are a problem that must be tackled if the numerous ground plants are to flourish.

Wild Daffodils

A view of Little Wood in spring where I have introduced wild daffodils. There are just two problems: this, like some of Owl's Wood, is relatively dry so maybe not ideal for them. Also, I do not believe all the bulbs sold to me are true native species as I intended. The daffodils precede the carpets of wood anemones and bluebells.

Barn area

Arable land for several hundred years. Approximately 3 acres.

Strictly speaking, this does not belong to the woods, although there is a gully on the south side that must once have been part of Shaw Wood and therefore ancient. Each year the gully is transformed by solid banks of bluebells (unless the sheep get in) that are shaded by Hornbeam, Ash and Field Maple. The rest consists of a huge barn built in the 1960s. The land is mainly grazed by my handful of animals, and some chickens have their home there in a fenced-off area. However, it has an interesting feature: an old disused railway cutting (axed by Beeching in the 1960s) behind the barn.

On the banks of the cutting there are some quite mature trees – mainly Pendunculate Oak, Ash, Hawthorn and some conifers. There are also a number of wild dog roses on the banks (see photograph page 72). My guess is that most of these trees were just saplings when the railway was built and were left when it was dismantled. They are now, therefore, about a century and a half old. The old maps distinctly show that the area was agricultural prior to the railway.

Hawthorns are beautiful trees if allowed to grow freely as some do here – clothed white in spring and red in autumn. Most were planted in hedges, so one does not often see them in all their glory. They are long-lived but do not grow tall – only to about ten metres or so – and are totally steeped in legend and folklore. Many people still believe that the felling of a Hawthorn spells doom and disaster to the perpetrator. Woe betides anyone contemplating the destruction of those now mature trees growing in this cutting!

Railway cutting in spring

In this aerial view the railway cutting appears almost totally overgrown with trees; in reality there is still quite a large flat and open area in the centre.

The most famous legend attached to the Hawthorn is, of course, the one surrounding the Glastonbury or Holy Thorn (*Crataegus monogyna praecox* or *'Biflora'*) which flowers both at Christmas and Easter. I have always wanted to grow one, but somehow have never got round to it. It will not flower if grown from seed – only if a branch is grafted on to the Common Hawthorn. The latter is a true native, whereas the Glastonbury Thorn reputedly owes its existence to Joseph of Arimathea who, when landing on the Isle of Avalon, thrust his Hawthorn staff of Middle Eastern origin into the ground. A number of specimens grow outside Glastonbury, including one in the Oxford parks where there is a 'Thorn Walk' boasting over forty species of Hawthorn!

Apart from the trees in the railway cutting, there are also a few plants that grow here and, as far as I can make out, nowhere else. These include Lesser Burdock, which one would normally think of as a woodland plant. Here they grow in the open on almost pure clay. Unique to the cutting are also two species of Agrimony – some of which have now spread elsewhere – as well as Great Mullein and Red Bartsia. There is also a large quantity of the infamous Ragwort.

I have had donkeys, a handful of sheep and Shetland ponies grazing in this area, but none have touched the Ragwort, nor indeed any of the other wild flowers. The net result is that the grass is slowly disappearing under Creeping Buttercup, Dock, Nettle and Ragwort. It is a situation that we will have to reverse – difficult without chemicals!

Ponies

My Shetland ponies grazing in the railway cutting. There is very little grass – mostly Dock and Creeping Buttercup – which the ponies will not eat and which we are trying to control by constant cutting.

The Biodiversity

Trees, Shrubs and Climbers

N=native, I=introduced, *=ancient woodland indicator

Scientific name	Common name	N/I	Remarks
Abies veitchii	Veitch's Silver Fir	I	On border of Owl's Wood and garden
Acer campestre	Field Maple, Common Maple	N*	Widespread. See illustration p63.
Acer platanoides	Norway Maple	I	A few planted in Butler's Wood
Acer pseudoplatanus	Sycamore	I	Only one or two present
Acer saccharum	Sugar Maple	I	I planted a handful
Aesculus x carnea	Red Horse Chestnut	I	Along northern edge of Shaw Wood
Aesculus hippocastanum	Horse Chestnut	I	Along northern edge of Shaw Wood. Others have seeded elsewhere.
Alnus glutinosa	Alder, Black Alder	N	Widespread. See illustration p62.
Betula pendula	Silver Birch	N	Widespread. See illustration p62.
Betula pubescens	Downy Birch	N	Mainly Pope's Wood
Bryonia dioica	White Bryony	N	Widespread but not many present
Calystegia sepium	Hedge Bindweed	N	Mainly in garden but some in woods. See illustration p64.
Carpinus betulus	Hornbeam	N*	Widespread. See illustration p63.
Castanea sativa	Sweet Chestnut	I	Mainly in Butler's and Badgers' Woods
Cedrus atlantica	Atlas Cedar	I	Only one or two planted by myself
Cedrus deodara	Deodar Cedar	I	Only one or two planted by myself
Chamaecyparis lawsoniana	Lawson's Cypress	I	Very few in woods – left for nesting birds
Convolvulus arvensis	Field Bindweed	N	Mainly in garden but some in woods. See illustration p64.
Cornus sanguinea	Dogwood	N	A handful planted by me in Shaw and Butler's Woods
Corylus avellana	Hazel	N	Widespread. See illustration p65.
Cotoneaster frigidus	Himalayan Tree-Cotoneaster	I	One in Christmas Wood and one in Streake's Wood
Crataegus monogyna	Hawthorn	N	Widespread. See illustration p65.
Cytisus scoparius	Broom	N	Widespread. See illustration p67.
Euonymus europaeus	Spindle Tree	N	Some present before 1987 and many planted after. See illustration p66.
Fagus sylvatica	Common Beech	N	Widespread. See illustration p67.
Fagus sylvatica var.purpurea	Copper Beech	I	Only in Streake's Wood and one on border of Owl's Wood
Frangula alnus	Alder Buckthorn	N*	Present before 1987 but more planted in Owl's Wood. See illustration p69.
Fraxinus excelsior	Common Ash	N	Widespread. See illustration p68.
Gymnocladus dioica	Kentucky Coffee-Tree	I	One on border of Owl's Wood
Hedera helix agg.	Ivy	N	Widespread
Ilex aquifolium	Holly	N*	Widespread. See illustration p68.
Juglans nigra	Black Walnut	I	A handful planted in Owl's Wood
Larix kaempferi	Japanese Larch	I	One massive tree at edge of Christmas Wood
Larix x eurolepis	Hybrid Larch	I	A fair number in Morgan's Strip and at edge of Shaw Wood
Ligustrum ovalifolium	Garden Privet	I	Garden escapees
Ligustrum vulgare	Wild Privet	N	A few on boundary of Pope's Wood
Lonicera periclymenum	Honeysuckle	N	Widespread. See illustration p69.
Malus sylvestris	Crab Apple	N*	One present after 1987; others planted subsequently. See illustration p68.
Picea abies	Norway Spruce	I	Christmas Wood and a few in Streake's Wood
Picea omorika	Serbian Spruce	I	Only two – one in Christmas Wood and one in Owl's Wood
Picea pungens	Colorado Spruce	I	Only a few on borders of Owl's and Pope's Woods
Picea sitchensis	Sitka Spruce	I	Edge of Owl's Wood
Pinus nigra ssp nigra	Austrian Pine	I	One in Barn area
Pinus sylvestris	Scots Pine	N	Too many but most in Badgers' Wood and Morgan's Strip. See illustration p70.
Platanus x hispanica	London Plane	I	Only one on northern border of Pope's Wood
Populus alba	White Poplar	N	Three only on border of Shaw Wood introduced via cuttings
Populus balsamifera	Balsam Poplar	I	Cuttings in Shaw and Little Woods
Populus balsamifera var.	Balsam Spire Poplar	I	Widespread in Pope's, Badgers' and Owl's Woods and Morgan's Strip
Populus nigra	Black Poplar	N	Only four introduced but more planned for Pope's Wood
Populus tremula	Aspen	N*	At present four in Owl's Wood and several in Butler's Wood. See illustration p55.
Populus x canadensis	Black Poplar clone	I	Widespread but most are doomed
Prunus avium	Wild Cherry, Gean	N*	Widespread. See illustration p70.
Prunus laurocerasus	Laurel, Cherry Laurel	I	Widespread but now declining
Prunus padus	Bird Cherry, Hawkberry	N*	Only a handful planted recently
Prunus spinosa	Blackthorn	N	Widespread in Morgan's Strip
Pseudotsuga menziesii	Douglas Fir	I	A fair number that survived 1987 hurricane
Quercus ilex	Holm Oak	I	A few recently planted on border of Owl's Wood
Quercus robur	Pendunculate Oak	N	Widespread. See illustrations pp7 and 71.

Quercus rubra	American Oak	I	One on border of shrubbery and Shaw Wood
Quercus rubra 'aurea'	Golden Oak	I	One specimen in Little Wood
Rhododendron ponticum	Rhododendron	I	Widespread but now declining
Ribes sylvestris	Redcurrant	N	See illustration p73
Rosa arvensis	Field Rose	N*	Widespread
Rosa canina	Dog Rose	N	Widespread. See illustration p72.
Rosa rubiginosa	Sweet-briar	N	Only one recently planted
Rosa spp	Wild Rose hybrids	N	Widespread
Rubus caesius	Dewberry	N	Fairly widespread
Rubus fruticosus	Bramble, Blackberry	N	Widespread
Rubus idaeus	Raspberry	I	Pope's Wood
Salix alba	White Willow	N	A few in Pope's Wood
Salix caprea	Goat Willow	N	Widespread. See illustration p72.
Salix cinerea ssp oleifolia	Grey Willow	N	Pope's Wood
Salix fragilis	Crack Willow	N	Some in Pope's Wood
Sambucus nigra	Elderberry	N	Widespread
Sequoia sempervirens	Coast Redwood, Wellingtonia	I	Only one on border of Owl's Wood
Sequoiadendron giganteum	Giant Redwood	I	A handful planted on borders of Owl's and Pope's Woods
Solanum dulcamara	Woody Nightshade, Bittersweet	N	Fairly widespread
Sorbus aria	Whitebeam	N	Widely introduced in Butler's Wood and elsewhere after The Storm. See illustration p73.
Sorbus aucuparia	Rowan, Mountain Ash	N*	Widespread
Sorbus torminalis	Wild Service Tree	N*	Many introduced in Butler's Wood
Symphoricarpos albus syn.	Snowberry	I	A few scattered specimens
Tamus communis	Black Bryony	N*	Widespread but not many present. See illustration p75.
Taxus baccata	Common Yew	N	Widespread. See illustration p74.
Tilia cordata	Small-leaved Lime	N*	A few introduced in Butler's Wood
Tilia x europaeus	Lime	I	A few mature trees
Ulex europaeus	Gorse	N	A solitary specimen in Shaw's Wood
Ulmus procera	English Elm	N	A few in Butler's Wood. Some introduced into Owl's Wood.
Viburnum lantana	Wayfaring Tree	N*	One or two present and more introduced recently
Viburnum opulus	Guelder Rose	N*	Widespread. See illustration p75.

I have included climbers in this survey, which is, strictly speaking, incorrect. Botanists normally divide these groups. However, I consider most climbers in a woodland to be part of the 'overstorey' as opposed to the 'understorey' and have to be managed as such. After all, many climbers, such as Bramble, White Bryony and Honeysuckle are powerful plants capable of strangling shrubs and young trees. It was also a good way to avoid too many lists.

When I first acquired the woods (nearly twenty-five years ago at the time of writing), more than half the forty acres were under mono-silviculture and the number of conifers far outstripped any broadleaved native species. This imbalance has been totally reversed. Only in Christmas Wood, Badgers' Wood, and to some extent in the Barn area are there still conifers in numbers. Most of the Scots Pine in Morgan's Strip are now gone. All the other woods now contain almost exclusively native species. There are only between thirty and forty native tree species in the UK (depending on which criteria you use for native) and almost double that number of shrubs.

We lack just five native tree species, four shrubs and one or two climbers. A few native species that were not present before The Storm were planted including some Small-leaved Lime, quite a number of which are thriving despite the fact that they prefer calcareous soil. Native trees or plants that were introduced by me are indicated under 'remarks' in the surveys. I cannot vouch for what might have been introduced by previous owners.

Perhaps one day I shall plant the missing species of native trees and shrubs? Someone is bound to accuse me of trying to turn the woods into an indigenous arboretum. Well, maybe, but I see nothing wrong in that if they thrive with no special treatment or attention. In fact, none of the trees and shrubs planted after The Storm was given undue special attention – there was just not enough time. All (approximately 4,000) had tree guards, without which most would not have survived the browsing of rabbits and deer. The guards have now all been removed, and many trees are now seed-bearing and self-generating.

I must say I am disappointed in some of the trees that were planted, especially the oaks. I do not believe the seedlings came from true native stock. And even if they did there are marked differences between them and the indigenous ones that survived the tempest. Also, the odd Small-leaved Lime has turned out to be the Common lime hybrid *Tilia x europaea!* There was so much hysteria and urgency to replant after The Storm that little attention was given to where the trees might have originated. However, as many of the self-seeded trees and shrubs are outstripping those that were planted, it may not make much difference in the long run.

Alder *Alnus glutinosa*

Host and habitat for hundreds! This is one of the many Alder clumps found in Pope's Wood where the majority of our Alders grow, die and are reborn. In legends, the Alder was sometimes called a 'faerie tree', allowing access to the realms of fairies; it is not hard to imagine elf-like creatures gambolling round the trunks. Siskins and Redpolls are attracted to Alders because of their nut-like fruits but, despite the fact that we have so many of this species we have only found Siskins and no Redpolls.

Silver Birch *Betula pendula*

Dainty, yet tough and weather-hardy and host to a wealth of wildlife, the Silver Birch has been called the 'Lady of the Woods'. It normally lives only about sixty years, but these 'twins' with their wonderful fissured barks were probably older. Alas, they recently succumbed to an excess of wind and water. They grew on the border between Owl's Wood and Pope's Wood.

Field Maple *Acer campestre*

This is a handsome, elegant native tree that displays a gamut of hues throughout the seasons and is quite glorious in autumn. It seeds itself and grows in all the woods except Pope's, but is only slowly self-generating in Badgers' Wood. The bark of mature trees is like a pale chequer-board, mossy and lichen-covered. The picture is of a mature specimen growing in Shaw Wood.

Hornbeam *Carpinus betula*

When we speak of catkins, the tendency is to think mainly of Hazel and Goat Willow, both of which produce a wonderful display in early spring. The Hornbeam flowers a little later and its catkins can be an arresting sight. They are often high in the canopy and therefore cannot be seen as easily as those of Hazel, which abound in hedgerows and spinneys. The Hornbeam germinates almost anywhere but has been much slower in spreading to the woods that were once solely conifers – mainly Badgers' Wood, Owl's Wood and Morgan's Strip. A mature Hornbeam is a handsome tree but, alas, no longer has any commercial value. I found a book about trees that does not even mention it! It is a pity considering how well this tree served communities for centuries, supplying indispensable fuel and charcoal for all kinds of purposes.

Hedge Bindweed *Calystegia sepium* and Field Bindweed *Convolvulus arvensis*

Our ancestors named both these climbers simply 'bindweeds', and I must say I am still slightly confused as to why two seemingly related plants should have totally different scientific names. In gardens, Bindweeds are a nuisance (both species are rampant in my vegetable garden), but in the woods they are comparatively rare – maybe because browsing animals love them (we give them, roots and all, to the sheep and ponies rather than risk them in the compost heaps). Yet both species look lovely scrambling up shrubs and hedges. We are planning a native hedge into which we should have no trouble introducing them together with Old Man's Beard (Clematis vitalba)*, which at present is absent in the woods.*

Hazel *Corylus avellana*

Hazel catkins can be an invigorating yellow/green sight even in mid-winter before anything else has yet stirred. Look at them closely some time and discover the tiny red female flowers which are often missed – overpowered by the flamboyant males. Most of our ancient native trees abound in folklore, and Shakespeare was often quick to use it in his plays (the squirrel he refers to is, of course, the now rare red squirrel as the grey did not then live in the UK):

> *Her chariot is an empty hazelnut,*
> *Made by the joiner squirrel, or old grub,*
> *Time out of mind the fairies' coach-makers.*

Squirrels are not the only ones that strip the nuts off Hazel: the nuts are one of the dormouse's favourite foods, and we can often find nut shells bearing the distinct teeth-marks of this much more welcome rodent.

Hawthorn *Crataegus monogyna* (see also photograph page 58)

Hawthorn grows in almost all the woods, with the exception of Pope's Wood. Hawthorn would not like the boggy soil there. They germinate freely everywhere else, which is sometimes rather galling. Have you ever tried growing them from seed? If so, you will know how difficult and temperamental they are! If allowed to grow into a mature tree, they make a lovely sight and are, of course, wonderful for wildlife. The blossoms are normally white, but are occasionally tinged pink. The Midland Hawthorn (C. laevigata*), which does not grow in the south of England, is always pink.*

Spindle Tree ***Euonymus europaeus***

Spindles were always present in the woods but hard to find after The Storm. A number have been planted, especially in New Wood and Butler's Wood. I wish the Spindle Tree were planted much more, both in gardens and in the wild. They seem to have everything – they are not too large, have intriguing multicoloured and faceted trunks and branches, stunning autumn colours and birds adore them. Birds peck out the seeds immediately the fruits split. Unfortunately, deer and various rodents also have a special taste for them. Spindle will regrow, but only as a coppiced bush. I have had to protect those that I would like to grow into trees.

Beech *Fagus sylvatica*

The oldest beeches are in Shaw Wood, but even there many of the largest specimens were lost in The Storm. Both planted and self-sown Beech grow in all the woods. In Butler's Wood there is the unusual sight of an enormous Beech shrub – a blown down Beech that has sprouted from a large trunk close to the ground. Beech were often pollarded but never coppiced in the past. In good years, the flowers can make quite a show, and there is little to rival their fresh spring leaves. Their autumn colours can also be arresting. This Beech grows on the edge of the pond in Shaw Wood and at sunset looked as if it were on fire. Unfortunately, a number of beeches in the wood are suffering from dieback, cause unknown.

Broom *Cytisus scoparius*

I did not see Broom here until about 1990 when The Storm had made clearings everywhere. At that time there were hardly any rides. Now Broom appears regularly along these rides and anywhere if it can get sufficient light. I have no idea whether the dormouse was present before The Storm and before I made all the changes in the woods, but the present structure, with plenty of rides along which Broom and Hazelnut grow, is certainly much more conducive for these little creatures. Ideally dormice like to jump from shrub to shrub, rather than run along the ground, and Broom is much to their liking.

Ash *Fraxinus excelsior*

Fortunately, Ash is widespread in all the woods and is self-seeding in large numbers. Their light canopy is welcome, enabling ground plants to obtain sufficient light for flowering whilst Beech, Oak and Hornbeam often shade them out. The largest Ash are found in the less boggy parts of Pope's Wood. Some have trunks three or four metres in circumference. The picture shows the male flowers looking a bit like some new curly hairstyle, and neither they nor the later-appearing females are normally seen, since they are often inconspicuous high in the canopy. The Beech and the Pendunculate Oak are sometimes called the 'Queen of the Forest' and 'King of the Forest'. If that is so, the Ash is surely the 'Princess' – gracious and elegant. Its leaves are often the last to appear and the first to fall. We have all heard some of the rhymes about the Oak and Ash; here is one from Surrey (I cannot find one from Sussex!):

If the Oak comes out before the Ash,
'Twill be a year of mix and splash
If the Ash comes out before the Oak,
'Twill be a year of fire and smoke.

Crab Apple *Malus sylvestris*

We definitely have one or two true native Crab Apple. Many were planted after The Storm in New Wood and Butler's Wood, but I doubt whether they were all true natives. These produce spiky thorns, which most of the planted ones seem not to possess. There was such a frenzy of planting after The Storm that many seedlings were sold as native but probably originated from anywhere but the UK. Nevertheless, the fallen, mushy apples of any species make excellent bird food.

Holly *Ilex aquifolium*

Both Holly and Ivy feature extensively in folklore. Holly is quite rampant in some parts of the woods, to the extent that we have cleared it in small areas. Under the canopy of large trees it seldom flowers or bears fruit. Normally male and female flowers are on separate trees (I try to preserve more females) and the picture is of the less flamboyant females showing the early signs of the eventual arresting berries. We have a few mature female trees. In some years they sparkle with an abundance of berries that are liable to disappear overnight, plundered by hungry birds.

Honeysuckle *Lonicera periclymenum*

Honeysuckle, or Woodbine, scrambling up a tree mature enough to hold its own (Honeysuckle can strangle saplings) is a beguiling sight. It grows everywhere but, like Ivy, it is inclined also to trail along shady woodland ground, never flowering and often ousting smaller plants. Honeysuckle nectar is food for many insects, including moths and butterflies. The scent is at its most powerful at night so moths especially tend to flock to it – it is said that moths can detect the scent a mile away. Books and people often refer to how good this plant is for bees. I have occasionally seen bees on Honeysuckle but never in large numbers.

Alder Buckthorn *Frangula alnus*

As far as I was aware, no Buckthorn grew here until some were planted in Owl's Wood after The Storm. In their first year they were decimated by the larvae of the Brimstone butterfly, but during the next ten to twelve years they grew into strong, healthy shrubs. However, in 2003 I discovered extensive damage to some of the main stems – the bark almost totally gnawed off. The damage was fairly high up so deer or grey squirrels seem the likely culprits, but it is odd they should have been left alone for so many years. But also in 2003 the shrub was found growing on the borders of Pope's Wood and Butler's Wood. It is highly unlikely they came from the seeds of the planted ones. The flowers are fairly inconspicuous, but the berries can be seen in several colours in autumn.

Scots Pine *Pinus sylvestris*

We have a surfeit of Scots Pine (see Badgers' Wood and Morgan's Strip), but as they were planted for commercial purposes, few, if any, display the stalwart spreading characteristics of trees growing naturally in the wild. The commercially grown trees are boringly straight and have few low branches as they are rarely given sufficient space and light. Male and female flowers are on the same tree. As shown in the picture they are bright and spectacular, often wafting clouds of pollen in the breeze.

Wild Cherry *Prunus avium*

Wild Cherry or Gean can grow enormous. A few of these giants were present before The Storm and they are greatly missed. Large numbers were subsequently planted, but we need not have bothered as seedlings have sprung up everywhere. Cherry blooms are short-lived but a magnificent sight, soon followed by a midsummer birds' feast, the birds reveling in the bitter fruit. Cherries can be prone to bad attacks of aphids but these in turn attract numbers of other insects including wasps and hover-flies which feed them to their larvae. Small birds also pick the aphids off the leaves.

Penndunculate Oak *Quercus robur*

This Oak grows at the top of Butler's Wood and is probably one of the oldest remaining trees – maybe two hundred or more years old. The virtues of the Pendunculate Oak as a provider of food and habitat for hundreds of creatures are well known. The Pendunculate Oak is in all the woods but, as far as I am aware, there is not a single Sessile Oak.

Goat Willow *Salix caprea*

Goat Willow or Sallow is the most widespread of all the Willow family in these woods. Everyone is familiar with the male flowers: the silky grey pussy willows that turn into bright yellow beacons in early spring. Pictured are the cone-like green female flowers which are less conspicuous. In a matter of days, they turn into white fluff that is wafted everywhere, ensuring wide germination. In the past, willows were one of the most important trees in cottage industries – they had myriad uses: baskets, thatching, sailing boats, hats, fodder for livestock and much else. Also, one should not forget that witches' broomsticks were often made from Willow. They also feed and house about 450 species of insects, and birds and mammals use them too.

Dog Rose *Rosa canina*

I must apologise for this bad picture, but I could not resist showing the amazing sight of a Dog Rose some ten metres up in the canopy of an oak tree. This one grows on the steep bank of the railway cutting in the Barn area where it gets plenty of light on one side. I tried to take another picture in better weather conditions but a hail storm had intervened and knocked off every single petal! Dog Roses grow more commonly in hedgerows, but the way hedges are often butchered these days in early autumn (quite the wrong time to cut hedges) they are not often seen like this. A neighbouring organic farm has allowed the hedges to grow as they should and wonderful arching sprays of roses can now be seen in late spring/early summer.

Redcurrant *Ribes sylvestris*

Large numbers of these currants grow in Streake's Wood and Pope's Wood, and I can only assume they are native, unless some farmer grew them commercially nearby a hundred or more years ago. They seem to have few fruit (unless the birds get at them before I can observe them!) and I found the specimen pictured in very wet ground almost totally hidden by Hemlock Water-dropwort and sedges.

Whitebeam *Sorbus aria*

I do not think any Whitebeam were present in the woods prior to The Storm. A number were planted on the slopes of Butler's Wood, where they thrive. Their wavy silver foliage, blooms and berries make a welcome addition to the biodiversity of the woods. Birds love the berries of all Sorbus *species, and this one is no exception. There are numerous local races of Whitebeam ranging from the south of England to the far north of Scotland. Some are said to be hybrids between this and the Wild Service Tree (*Sorbus torminalis*). As both are now growing well in Butler's Wood, who knows, we might spawn a new hybrid!*

Yew *Taxus baccata*

There are only three native conifers: Scots Pine, Juniper and Yew. Like so many native trees, the Yew is steeped in tradition and folklore, and was revered by the Celts, Romans and early Christians alike. As far as wildlife is concerned, birds relish the soft outer flesh of the berries, 'spitting out' the poisonous seed, thus promoting widespread germination. It provides dense cover all the year, and sometimes birds such as Thrushes nest in the lower branches. Yew is found everywhere in the woods, but as it only grows very slowly, small seedlings are often overcome by other plants. Our largest Yew, pictured below, grows on the borders of Christmas Wood and Pope's Wood and, although large, it cannot be considered old in Yew terms – it is maybe a hundred years or more.

Guelder Rose *Viburnum opulus*

This lovely shrub is one of my favourites. Unfortunately, it is also much favoured by deer and small rodents. However, it does coppice well, so individual plants are rarely destroyed. It looks good in spring with its 'clever' flowers (the showy white outer ones are sterile and only there to attract insects to fertilize the insignificant inner ones), but it really comes into its own in autumn with its stunning berries and foliage. It grows throughout the woods. Some were planted after The Storm, especially in New Wood and Owl's Wood, but many were already present elsewhere. Some people tell me the berries are poisonous: this is not true, although they are extremely bitter. Maybe this is why the berries are the last to go in winter – the birds do not find them particularly palatable. In Canada they are sometimes used as a substitute for cranberries. The shrub should be flattered – I do not think there are many plants with so many alternative names: Cramp Bark, Snowball Tree, King's Crown, High Cranberry, Red Elder, Rose Elder, Water Elder, May Rose, Whitsun Rose, Dog Rowan Tree, Silver Bells, Whitsun Bosses, Gaitre Berries and Black Haw.

Black Bryony *Tamus communis*

This climber, like Honeysuckle and Ivy, can strangle small trees and shrubs but, twining up large trees, they make a pretty sight with their bright berries, especially in deep winter when all else seems bleak.

All parts of this plant, including the roots, are poisonous; I feel sure it must have a useful purpose to wildlife somehow, though I have yet to find it. It has been used by humans as a powerful diuretic.

Plants

A=annual, Bi=biennial, P=perennial, B=bulb, O=orchid, N=native, I=introduced
*denotes ancient woodland indicator

Scientific name	**Common name**	**Type**	**N/I**	**Comments**
Achillea millefolium	Yarrow	P	N	Infrequent except in garden
Adoxa moschatellina	Moschatel, Town Hall Clock	P	N*	See illustration p82
Aegopodium podagraria	Ground Elder, Bishop's Weed	P	N	Widespread, especially in garden!
Aethusa cynapium	Fool's Parsley	A	N	Infrequent
Agrimonia eupatoria	Agrimony	P	N	Mostly in Barn area
Agrimonia procera	Fragrant Agrimony	P	N	Barn area
Agrostemma githago	Corncockle	A	I	See illustration p82
Agrostis capillaris	Common Bent-grass	P	N	Widespread
Ajuga reptans	Bugle	P	N	Widespread
Alisma plantago-aquatica	Water-plantain	P	N	Only appeared in vegetable garden pond
Allium ursinum	Ramsons	P	N*	Some introduced to Pope's Wood
Alopecurus pratensis	Meadow Foxtail	P	N	Widespread in open places
Anagallis arvensis	Scarlet Pimpernel	A	N	See illustration p83
Anemone nemorosa	Wood Anemone	P	N*	Widespread in ancient woodland
Angelica sylvestris	Wild Angelica	P	N	Mainly in Pope's Wood and Shaw Wood
Anthemis arvensis	Corn Chamomile	A	N	Only in our 'cornfield'
Anthemis cupriana	Chamomile	P	I	A few introduced by me
Anthemis tinctoria	Yellow Chamomile	P	I	A few introduced by me
Anthoxanthum odoratum	Sweet Vernal-grass	P	N	Widespread
Anthriscus sylvestris	Cow Parsley	P	N	Mainly along drive to house
Arabidopsis thaliana	Thale Cress	P	N	Widespread
Arctium minus	Lesser Burdock	P	N	See illustration p84
Arum italicum	Italian Lords and Ladies	P	N	Not as frequent as *A. maculatum*
Arum maculatum	Lords and Ladies	P	N	Widespread throughout all the woods. See illustration p83.
Athyrium filix-femina	Lady Fern	P	N	See illustration p82
Bellis perennis	Common Daisy	P	N	On rides
Blechnum spicant	Hard Fern	P	N*	Frequent on wet banks. See illustration p87.
Borago officinalis	Borage	A	N	Escapee from herb garden
Brachypodium sylvaticum	False Brome	P	N	Widespread
Bromopsis ramosa	Hairy Brome	A	N	Widespread
Callitriche stagnalis	Common Water-starwort	P	N	See illustration p85
Caltha palustris	Marsh Marigold	P	N	See illustration p85
Campanula trachelium	Nettled-leaved Bellflower	P	N	See illustration p86
Capsella bursa-pastoris	Shepherd's Purse	A	N	Widespread but more common in vegetable garden
Cardamine flexuosa	Wavy Bitter-cress	Bi	N	Widespread
Cardamine pratensis	Cuckoo Flower, Lady's Smock	P	N	Widespread
Carex echinata	Star Sedge	P	N	Pope's Wood
Carex flacca	Glaucous Sedge	P	N	Widespread
Carex hirta	Hairy Sedge	P	N	Mainly in Pope's Wood
Carex laevigata	Smooth-stalked Sedge	P	N*	Mainly in Pope's Wood
Carex ovalis	Oval Sedge	P	N	Mainly in Pope's Wood
Carex pallescens	Pale Sedge	P	N*	Mainly in Pope's Wood
Carex paniculata	Greater Tussock Sedge	P	N	See illustration p32–33.
Carex pendula	Pendulous Sedge	P	N*	See illustration p86.
Carex pilulifera	Pill Sedge	P	N	Mainly in Pope's Wood
Carex remota	Remote Sedge	P	N*	Mainly in Pope's Wood
Carex sylvatica	Wood Sedge	P	N*	Mainly in Pope's Wood
Carex vesicaria	Bladder-sedge	P	N	Mainly in Pope's Wood
Centaurea cyanus	Cornflower	A	N	Only present in our 'cornfield'
Centaurea nigra	Black Knapweed	P	N	Widespread
Centaurium erythraea	Common Centaury	A	N	Widespread on rides
Centaurium pulchellum	Lesser Centaury	A	N	Widespread on rides
Cerastium fontanum	Common Mouse-ear	P	N	Widespread on rides
Cerastium glomeratum	Sticky Mouse-ear	A	N	Widespread on rides
Chamerion angustifolium	Rosebay Willowherb	A	N	Becoming a nuisance in sunny areas
Chenopodium album	Fat-hen	A	N	Usually near cultivated areas

A=annual, Bi=biennial, P=perennial, B=bulb, O=orchid, N=native, I=introduced

*denotes ancient woodland indicator*denotes ancient woodland indicator

Scientific name	Common name	Type	N/I	Comments
Chenopodium ficifolium	Fig-leaved Goosefoot	A	N	Usually near cultivated areas
Chenopodium polyspermum	Many-seeded Goosefoot	A	N	Usually near cultivated areas
Chrysanthemum leucanthemum	Oxeye Daisy	P	N	As yet infrequent outside garden area
Chrysanthemum segetum	Corn Marigold	A	N	See illustration p87
Chrysosplenium oppositifolium	Opposite-leaved Golden Saxifrage	P	N*	See illustration p87
Circaea lutetiana	Enchanter's Nightshade	P	N	See illustration p88
Cirsium arvense	Creeping Thistle	P	N	Widespread
Cirsium palustre	Marsh Thistle	Bi	N	See illustration p88
Cirsium vulgare	Spear Thistle	Bi	N	Widespread
Conopodium majus	Pignut	P	N*	Widespread. See illustration p86
Coronopus didymus	Lesser Swine-cress	A	I	Widespread
Cyclamen hederifolium	Cyclamen, Sowbread	P	I	A few in Owl's Wood self sown from garden
Cynosurus cristatus	Crested Dog's-tail	P	N	See illustration p84
Dactylis glomerata	Cocks-foot	P	N	Rides
Dactylorhiza fuchsii	Common Spotted Orchid	O	N	Widespread except Badgers' Wood and Owl's Wood
Dactylorhiza praetermissa	Southern Marsh Orchid	O	N	Introduced by me in 2003
Daucus carota	Wild Carrot	A	N	See illustration p88
Deschampsia flexuosa	Wavy-hair Grass	P	N	See illustration p89
Digitalis purpurea	Foxglove	Bi	N	See illustration p90
Dipsacus fullonum	Wild Teasel	Bi	N	See illustration p91
Dryopteris affinis agg.	Scaly Male Fern	P	N*	Widespread in the ancient woods
Dryopteris carthusiana	Narrow Buckler Fern	P	N	Mainly in Pope's Wood
Dryopteris dilatata	Broad Buckler Fern	P	N	Widespread
Dryopteris felix-mas	Male Fern	P	N	Widespread
Elytrigia repens	Common Couch	P	N	Widespread
Epilobium ciliatum	American Willowherb	P	I	Widespread
Epilobium hirsutum	Greater Willowherb	P	N	Widespread in wetter areas
Epilobium montanum	Broad-leaved Willowherb	P	N	Widespread in damp places
Epipactis helleborine	Broad-leaved Helleborine	O	N*	See illustration p91
Epipactis palustris	Marsh Helleborine	O	N	Introduced by me in 2003
Equisetum arvense	Field Horsetail	P	N	Widespread
Erythronium dens-cani	Dog's-tooth Violet	P	I	See illustration p91
Eupatorium cannabinum	Hemp Agrimony	P	N	See illustration p92
Euphorbia helioscopia	Sun Spurge	A	N	Occasional
Euphorbia peplus	Petty Spurge	A	N	Occasional
Euphrasia nemorosa	Common Eyebright	A	N	See illustration p92
Festuca gigantea	Giant Fescue	P	N*	Fairly widespread
Festuca rubra agg.	Red Fescue	P	N	Damp places
Fragaria vesca	Wild Strawberry	P	N	Widespread in ancient woodland
Galanthus nivalis	Snowdrop	B	I	Recently introduced by me into Shaw Wood
Galeopsis tetrahit agg.	Common Hemp Nettle	A	N	Mainly Barn area
Galium aparine	Cleavers	A	N	Widespread
Galium palustre agg.	Common Marsh-bedstraw	P	N	Widespread
Galium saxatile	Heath Bedstraw	P	N	Widespread
Geranium dissectum	Cut-leaved Cranesbill	A	N	Fairly widespread
Geranium molle	Doves-foot Cranesbill	A	N	Infrequent
Geranium pratense	Meadow Cranes-bill	P	N	Occasional
Geranium robertianum	Herb Robert	A	N	See illustration p92
Geranium sanguineum	Bloody Cranesbill	P	N	See illustration p93
Geranium versicolor	Pencilled Cranesbill	P	I	Appears intermittently. Garden escapee.
Geum urbanum	Wood Avens	P	N	Widespread
Gladiolus illyricus ssp. byzantinus	Gladiolus	P	N	See illustration p94
Glechoma hederacea	Ground Ivy	P	N	See illustration p93
Glyceria fluitans	Floating Sweet-grass	P	N	Widespread
Gnaphalium uliginosum	Marsh Cudweed	A	N	Pope's Wood
Heracleum sphondylium	Hogweed	P	N	Widespread. See illustration p94.
Hieracium officinalis agg.	Hawkweed	P	N	Frequent in open ground
Holcus lanatus	Yorkshire Fog	P	N	Widespread in rides
Holcus mollis	Creeping Soft-grass	P	N*	Widespread in open places
Hyacinthoides non-scripta	Bluebell	B	N*	Dense on ancient woodland sites (about 25 acres)
Hydrocotyle vulgaris	Marsh Pennywort	P	N	Mainly in Pope's Wood

A=annual, Bi=biennial, P=perennial, B=bulb, O=orchid, N=native, I=introduced
*denotes ancient woodland indicator

Scientific name	Common name	Type	N/I	Comments
Hypericum humifusum	Trailing St. John's-wort	P	N	Colonies mainly in glade in Shaw Wood
Hypericum perforatum	Perforate St. John's-wort	P	N	Colonies mainly in glade in Shaw Wood
Hypericum pulchrum	Slender St. John's-wort	P	N*	See illustration p95
Hypericum tetrapterum	Square-stalked St. John's-wort	P	N	Colonies mainly in glade in Shaw Wood
Hypochaeris radicata	Cat's-ear	P	N	Common in dry grass areas
Iris pseudacorus	Yellow Iris	P	N	See illustration p95
Juncus articulatus	Jointed Rush	P	N	Mainly in Pope's Wood
Juncus bulbosus	Bulbous Rush	P	N	Mainly in Pope's Wood
Juncus conglomeratus	Compact Rush	P	N	Fairly widespread
Juncus effusus	Soft Rush	P	N	See illustration p95
Lamiastrum galeobdolon	Yellow Archangel	P	N*	Increasing in Pope's, Streake's and Owl's Woods
Lamium album	White Dead-nettle	P	N	Infrequent, mostly in glade in Owl's Wood
Lamium maculatum	Spotted Dead-nettle	P	I	Usually appears at the edge of Owl's Wood
Lamium purpureum	Red Dead-nettle	P	N	See illustration p96
Lapsana communis	Nipplewort	A	N	Frequent in open ground
Lathyrus pratensis	Meadow Vetchling	P	N	Mainly glade in Shaw Wood
Lemna minor	Common Duckweed	A	N	Frequent in water but never in Shaw Wood pond
Leontodon autumnalis	Autumn Hawkbit	P	N	Frequent in open ground
Leucanthemum vulgare	Oxeye Daisy	P	N	Increasing in grassy areas
Leucojum aestivum	Summer Snowflake	B	N	A few introduced by me
Linaria vulgaris	Common Toadflax	P	N	Mainly Barn area
Listera ovata	Common Twayblade	O	N	See illustration p96
Lolium perenne	Perennial Rye-grass	P	N	Widespread
Lotus corniculatus	Common Bird's-foot-trefoil	P	N	See illustration p93
Lotus pedunculatus	Greater Bird's-foot-trefoil	P	N	Frequent but mainly in glade in Shaw Wood
Luzula campestris	Field Wood-rush	P	N	Fairly widespread especially around Shaw Wood
Luzula multiflora	Heath Wood-rush	P	N	Fairly widespread especially around Shaw Wood
Luzula multiflora var.congresta	Heath Wood-rush	P	N	Mainly in Pope's and Shaw Woods
Luzula pilosa	Hairy Wood-rush	P	N*	Fairly widespread especially around Shaw Wood
Luzula sylvatica	Great Wood-rush	P	N*	Widespread on ancient woodland sites
Lychnis flos-cuculi	Ragged Robin	A	N	See illustration p97
Lycopus europaeus	Gypsywort	P	N	Barn area
Lysimachia nemorum	Yellow Pimpernel	P	N*	Widespread on rides
Lysimachia nummularia	Creeping-jenny	P	N	Widespread on rides
Lysimachia punctata	Dotted Loosestrife	P	I	On bank in Shaw Wood
Lysimachia vulgaris	Yellow Loosestrife	P	N	On bank in Shaw Wood
Lythrum salicaria	Purple Loosestrife	P	N	See illustration p98
Malva moschata	Musk Mallow	P	N	See illustration p98
Malva sylvestris	Common Mallow	P	N	Appears intermittently – garden escapee
Matricaria discoides	Pineappleweed	A	I	Barn area and chicken run
Matricaria recutita	Scented Mayweed,	P	N	Mainly Barn area
Medicago lupulina	Black Medick	P	N	Widespread
Melampyrum pratense	Common Cow-wheat	A	N*	See illustration p21
Melica uniflora	Wood Melick	P	N*	See illustration p99
Mentha aquatica	Water Mint	P	N	Widespread in Pope's Wood
Mentha arvensis	Corn Mint, Field Mint	P	N	Widespread
Menyanthes trifoliata	Bogbean	P	N	Introduced to pond in vegetable garden
Milium effusum	Wood Millet	P	N*	Fairly widespread
Mimulus guttatus	Monkey Flower	P	I	Only by pond in vegetable garden
Molina caerulea	Purple Moor-grass	P	N	Widespread
Myosotis arvensis	Field Forget-me-not	P	N	Fairly widespread in grass
Myosotis sylvatica	Wood Forget-me-not	P	N*	Widespread
Narcissus poeticus	Pheasant's Eye Daffodil	B	I	A few introduced by me into Little Wood. Not doing well.
Narcissus pseudonarcissus	Wild Daffodil, Lenten Lily	B	N	Introduced by me into Shaw Wood and Little Wood
Narcissus spp	Daffodil	B	I	Shaw Wood
Nuphar lutea	Yellow Water Lily	P	N	Pond in Barn area but always eaten by ducks
Odontites vernus	Red Bartsia	A	N	Railway cutting
Oenanthe crocata	Hemlock Water-dropwort	P	N	See illustration p34
Oenothera biennis	Common Evening Primrose	Bi	I	Garden escapee
Onopordum acanthium	Scotch Thistle, Cotton Thistle	Bi	N	Appears intermittently – probably garden escapee
Orchis mascula	Early Purple Orchid	O	N	See illustration p98

A=annual, Bi=biennial, P=perennial, B=bulb, O=orchid, N=native, I=introduced
*denotes ancient woodland indicator

Scientific name	Common name	Type	N/I	Comments
Oxalis acetosella	Wood Sorrel	P	N	See illustration p101
Oxalis articulata	Pink Sorrel	P	I	Garden escapee
Oxalis corniculata	Yellow Oxalis	P	I	Probably garden escapee
Papaver dubium	Long-headed Poppy	A	N	Appears in disturbed ground
Papaver lecoqii	Yellow-juiced Poppy	A	I	Appears in disturbed ground
Papaver rhoeas	Field Poppy, Corn Poppy	A	N	See illustration p100
Papaver somniferum	Opium Poppy	A	I	Cornfield on lawn and vegetable garden
Pentaglottis sempervirens	Green Alkanet	P	I	Garden escapee
Persicaria bistorta	Common Bistort	P	N	Fairly widespread
Persicaria hydropiper	Water-pepper	A	I	Mainly Pope's Wood
Persicaria maculata	Redshank	A	N	Widespread
Phyllitis scolopendrium	Hart's Tongue	P	N*	Only a handful present in Pope's Wood and Badgers' Wood
Pimpinella saxifraga	Burnet-saxifrage	P	N	Barn area
Plantago lanceolata	Ribwort Plantain	P	N	Widespread
Plantago major	Greater Plantain	P	N	Widespread
Poa annua	Annual Meadowgrass	A	N	Widespread on rides
Poa nemoralis	Wood Meadowgrass	P	N	Widespread
Poa trivialis	Rough Meadowgrass	P	N	Widespread
Polemonium caeruleum	Jacob's Ladder	P	N	A few introduced by me
Polygonatum multiflorum	Solomon's-seal	P	N	See illustration p101
Polygonum aviculare	Knotgrass	A	N	Fairly common
Polypodium vulgare	Common Polypody	P	N*	See illustration p101
Potamogeton natans	Broad-leaved Pondweed	P	N	Pond in Pope's Wood
Potamogeton polygonifolius	Bog Pondweed	P	N	Pond in Pope's Wood
Potentilla anserina	Silverweed	P	N	Occasional
Potentilla erecta	Tormentil	P	N	Widespread
Potentilla reptans	Creeping Cinquefoil	P	N	Widespread
Potentilla sterilis	Barren Strawberry	P	N*	Widespread
Primula elatior	Oxlip	P	N	Introduced by me
Primula veris	Cowslip	P	N	Introduced by me
Primula vulgaris	Primrose	P	N*	Widespread
Prunella vulgaris	Selfheal	P	N	Widespread
Pteridium aquilinum	Bracken	P	N	Widespread
Pulicaria dysenterica	Common Fleabane	P	N	Originally only in Barn area but now spreading
Ranunculus acris	Meadow Buttercup	P	N	Widespread
Ranunculus bulbosus	Bulbous Buttercup	P	N	Widespread
Ranunculus ficaria agg.	Lesser Celandine	P	N	Widespread
Ranunculus flammula	Lesser Spearwort	P	N	Common along wet rides and by streams. See illustration p102.
Ranunculus repens	Creeping Buttercup	P	N	Widespread
Rorippa nasturtium-aquaticum	Water-cress	P	N	By spring in Pope's Wood
Rumex acetosa	Common Sorrel	P	N	Widespread
Rumex acetosella	Sheep's Sorrel	P	N	Widespread
Rumex crispus ssp crispus	Curled Dock	P	N	Widespread
Rumex obtusifolius	Broad-leaved Dock	P	N	Widespread
Rumex sanguineus	Wood Dock, Blood-veined Dock	P	N	Widespread
Sagina apetala agg.	Annual Pearlwort	A	N	Widespread
Sagina procumbens	Procumbent Pearlwort	P	N	Widespread
Sanicula europaea	Sanicle	P	N*	See illustration p102
Scirpus sylvaticus	Wood Club-rush	P	N*	See illustration p97
Scrophularia auriculata	Water Figwort	P	N	Barn Area and Pope's Wood
Scrophularia nodosa	Common Figwort	P	N	Widespread
Scutellaria galericulata	Skullcap	P	N	Pope's Wood
Senecio erucifolius	Hoary Ragwort	Bi	N	Mainly Barn area but spreading
Senecio jacobaea	Common Ragwort	Bi	N	See illustration p154.
Senecio sylvaticus	Heath Groundsel	A	N	Fairly common
Senecio vulgaris	Groundsel	A	N	Widespread
Silene latifolia	White Campion	P	N	See illustration p102.
Silene dioica	Red Campion	P	N	Widespread
Silybum marinum	Milk Thistle	Bi	I	One appeared in 2002/3
Sinapis arvensis	Charlock	A	N	More common in garden

A=annual, Bi=biennial, P=perennial, B=bulb, O=orchid, N=native, I=introduced
*denotes ancient woodland indicator

Scientific name	Common name	Type	N/I	Comments
Sisymbrium officinale	Hedge Mustard	A	N	First appeared in Little Wood – spreading
Solanum nigrum	Black Nightshade	P	N	More common in garden
Soleirolia soleirolii	Mind-your-own-business	P	I	Garden escape
Solidago virgaurea	Goldenrod	P	N*	Large colonies in Shaw Wood. See illustration p103.
Sonchus arvensis	Perennial Sow Thistle	P	N	Widespread
Sonchus asper	Prickly Sow Thistle	A	N	Widespread
Sonchus oleraceus	Smooth Sow Thistle	A	N	Fairly common. See illustration p105.
Sparganium erectum	Branched Bur-reed	P	N	Only in Popes Wood
Spergula arvensis	Corn Spurrey	A	N	On disturbed ground
Stachys arvensis	Field Woundwort	A	N	Mainly Barn area
Stachys officinalis	Betony	P	N*	See illustration p104
Stachys sylvatica	Hedge Woundwort	P	N	See illustration p104
Stellaria graminea	Lesser Stitchwort	P	N	Widespread
Stellaria holostea	Greater Stitchwort	P	N	Widespread
Stellaria media	Common Chickweed	A	N	Widespread
Stellaria uliginosa	Bog Stitchwort	P	N	Fairly common in wet areas
Succisa pratensis	Devil's-bit Scabious	P	N	Occasional on rides
Symphytum x uplanbicum	Russian Comfrey	P	N	Garden escape
Tanacetum parthenium	Feverfew	P	N	Garden escape
Taraxacum officinale agg	Dandelion	P	N	Widespread
Teucrium scorodonia	Wood Sage	A	N	See illustration p105
Torilis japonica	Upright Hedge-parsley	A	N	Barn area
Tragopogon pratensis agg	Goat's-beard	Bi	N	Fairly widespread
Trifolium campestre	Hop Trefoil	A	N	Mainly in glades
Trifolium dubium	Lesser Trefoil	A	N	Mainly in glades
Trifolium micranthum	Slender Trefoil	A	N	Mainly in glades
Trifolium pratense	Red Clover	P	N	Widespread
Trifolium repens	White Clover	P	N	Widespread
Tripleurospermum inodorum	Scentless Mayweed	A	N	Widespread in Barn area
Tussilago farfara	Colt's-foot	P	N	Occasional
Typha latifolia	Reedmace, Bulrush	P	N	Appears sporadically in Pope's Wood pond
Urtica dioica	Common Nettle	P	N	Widespread
Valeriana officinalis	Common Valerian	P	N	Fairly widespread
Verbascum thapsus	Great Mullein	A	I	Only in Barn Area
Veronica beccabunga	Brooklime	P	N	Introduced into pond in vegetable garden
Veronica chamaedrys	Germander Speedwell	P	N	Widespread
Veronica montana	Wood Speedwell	P	N*	Widespread
Veronica officinalis	Heath Speedwell	P	N	Widespread
Veronica persica	Common Field Speedwell	P	N	Widespread
Veronica polita	Grey Field Speedwell	A	N	Widespread
Veronica serpyllifolia	Thyme-leaved Speedwell	P	N	Widespread
Vicia cracca	Tufted Vetch	P	N	Found intermittently on rides
Vicia hirsuta	Hairy Tare	A	N	Found intermittently on rides
Vicia sativa subsp. segetalis	Common Vetch	A	N	Found intermittently on rides
Vicia sativa subsp. nigra	Narrow-leaved Vetch	A	N	Found intermittently on rides
Vicia sepium	Bush Vetch	P	N*	Found intermittently on rides
Vicia tetrasperma	Smooth Tare	A	N	Found intermittently on rides
Viola arvensis	Field Pansy	A	N	Present some years ago, none recently
Viola odorata	Sweet Violet	P	N*	Widespread
Viola palustris	Marsh Violet	P	N*	Only in Pope's Wood
Viola reichenbachiana	Early Dog-violet	P	N	Widespread
Viola riviniana	Common Dog-violet	P	N	Widespread
Vulpia bromoides	Squirrel-tail Fescue	A	N	Found intermittently on rides

Opposite, Common Polypody *Polypodium vulgare*

See page 101.

I had a head start identifying the plants as the Sussex Botanical Recording Society (SBRS) had made a survey in June 1993. It included the trees and shrubs but not fungi, mosses or lichens and resulted in a formidable list of 250 species. However, it is an impossibility to find and record all the species in a few hours during one summer's day or, indeed, over several years. Many plants grow only sporadically, and others may suddenly appear out of nowhere. When I decided to write this book, I set about augmenting the SBRS list and photographing some of the species. I had absolutely no idea that I would end up with a staggering combined list of nearly 400! I have introduced a mere handful of native species that were not present in 1993, and I have also included a few that are really garden escapees, but I saw no reason to exclude them since most were native and had just decided to wander into the wild. I have indicated in the list those that I introduced and those that apparently appeared from nowhere.

It was immensely gratifying, having completed the list, to find we had no less than fifty-four ancient woodland indicators (including trees). We have another four species of ancient woodland indicators, but these were planted. Yet another could be a garden escapee; namely Solomon's-seal.

If you are thinking about growing wild species, be prepared for some surprises and many disappointments. Here are two of my experiences. Teasels are attractive to the eye and to birds; they reputedly like clay soil and grow in open woodland. In 1995, I sowed some Teasel seed in trays and planted them out at the edge of Little Wood in what appeared to be solid clay, hoping they would subsequently seed themselves. Not all survived, but a few flowered the following year. Nothing happened for several years until one giant specimen appeared at the edge of a farm field a considerable distance away (as far as I can make out this species grows nowhere else in the near vicinity). Again, nothing happened until 2002 when about four appeared in the shrubbery. The next year (eight years after the original planting) literally hundreds appeared where the original seedlings had been planted (see illustration, page 91). Birds, probably finches, plucked off nearly all the individual seeds.

I find all Campanulas appealing, and I attempted to naturalize the Nettled-leaved variety, despite the fact it prefers chalky soil. Again, I grew them in boxes and planted them out in Shaw Wood. Not one survived – rabbits ate them all down to the crown, so despite being perennials they had no chance to regrow. A second attempt in a different place was slightly more successful – just two survived to flower! It remains to be seen whether they will appear again and eventually seed themselves. Maybe I should have sown them directly into the ground to bear out my theory that naturally generated plants are more suited to resist the attacks of predators.

We recently 'tilled' part of the lawn and sowed cornfield flowers, some of which once adorned the whole countryside but are now becoming scarce. The results can be seen on page 107. Many would argue that neither they nor the field has a place in a book about a woodland. This book, however, is also about biodiversity, and since some of these plants are becoming a rarity (like ancient woodlands) I feel as much as possible should be done to preserve these beautiful flowers. Some would also contend that even with the wildest stretch of imagination these cannot be called 'natural'. True, but then what is wholly natural these days?

I still struggle to identify all the species that were recorded in 1993 (or indeed name some of the new ones), especially the grasses, sedges, rushes and others. However, as with the succeeding chapters, I have had the help of experts, all of whom have been wonderful, giving their help freely and enthusiastically. In this chapter it has been Paul Harmes especially (whose group was responsible for the original survey) who has come to my rescue. The following pages illustrate just a few of the species that now grow here.

Top left

Town Hall Clock ***Adoxa moschatellina***

This tiny, charming plant gets its name from the square of four florets (and one on top). It grows carpet-like everywhere in damp, shady places. Town Hall Clock is an ancient woodland indicator, and surprisingly there are large clumps in both New Wood and Streake's Wood. Are these the first signs that these woods are evolving into ancient ones?

Top right

Corncockle ***Agrostemma githago***

Corncockles are not classified as native but originate in Europe. They only grow in our 'created' meadow. They are rarely seen now since herbicides have eradicated them from traditional cornfields, along with Cornflower, Corn Marigold and others. The problem is that they only germinate on disturbed ground, so ploughed fields are ideal. Unfortunately, the seed is toxic to both humans and animals. Perhaps they will reappear on organic farms where, nevertheless, the seeds could still pose a problem.

Lady Fern ***Athyrium filix-femina***

The leaves of Ferns, even those of the pesky Bracken, are delicate and beautiful. But the fronds unfurling in spring have a majestic look, especially the large Lady Fern. Note the Early Purple Orchids on the left.

Scarlet Pimpernel *Anagallis arvensis*

Most people know this plant, and it is widespread along our rides. Next time you see one, look more closely at the wonderful colours and shading at its centre. The flower is normally less than 1cm wide. It is also called Poor Man's Weatherglass or Shepherd's Clock as it opens its petals at 8am and closes them at 2pm. In bad weather, they do not open at all. Scarlet Pimpernel also possesses a host of medicinal properties. If chickens consume it, they become euphoric and begin to laugh or chatter – hence the name Anagallis, *Greek for laugh. On the other hand, there have been reports of sheep dying after ingesting them.*

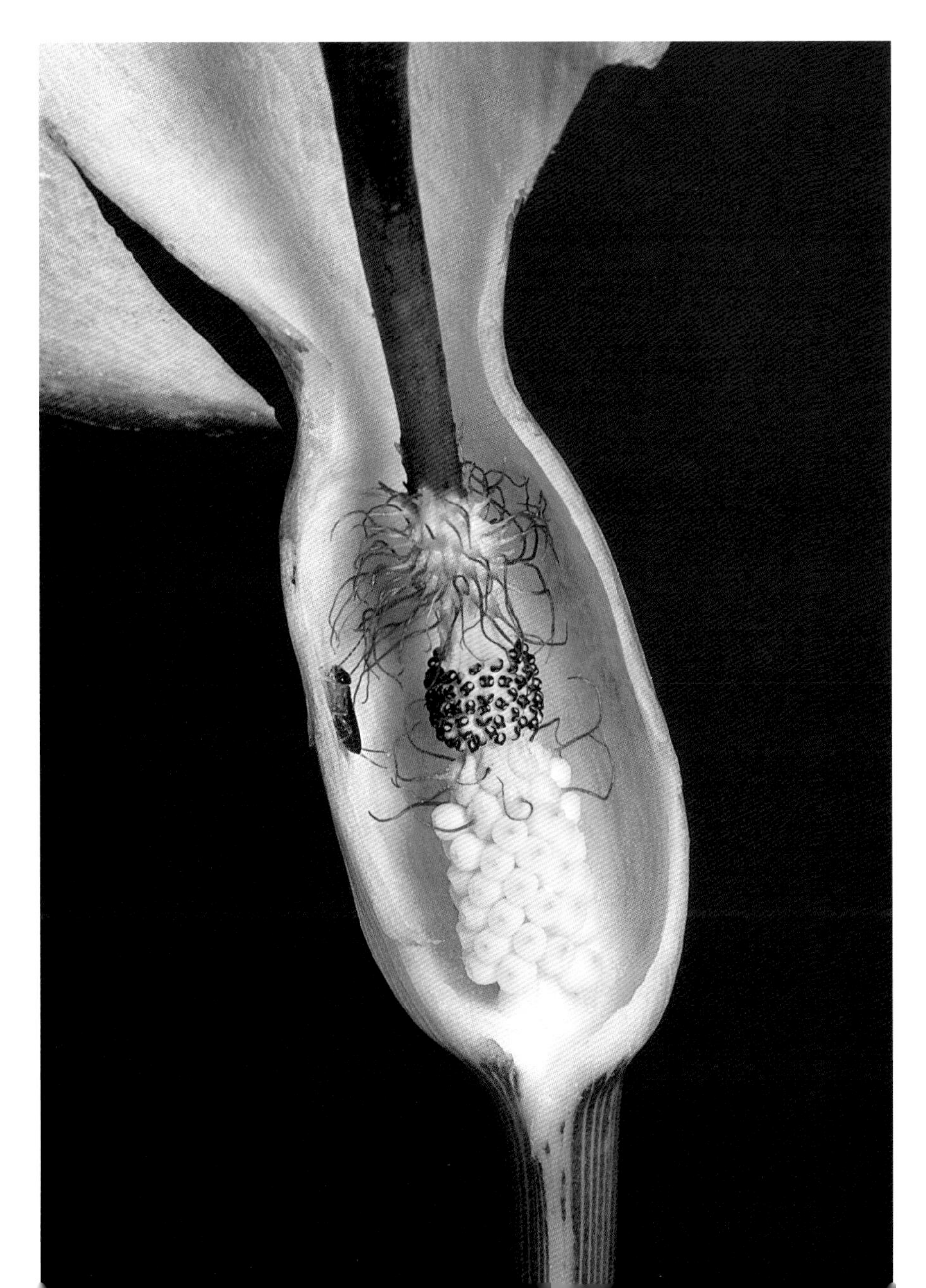

Lords and Ladies *Arum maculatum*

These, and their relative Arum italicum, *appear in all the woods. They are clever, devious plants. Neither the male nor female flowers are visible from the outside, so a scent lures insects into its 'cup', past bristly hairs that bend downwards but not easily upwards, to the distinctive female flowers at the bottom – passing the males on the way. This ensures pollination, and the hapless insect may later escape or drown in a sea of the mealy scent which attracted it in the first place! The results of all these machinations are bright red autumn berries that are poisonous.*

Crested Dog's-tail *Cynosurus cristatus*

Grasses should never be ignored. Many, if looked at in a certain light, become delightful architectural plants like this one. This is a particularly versatile grass: it grows in any soil, is very resistant to frost, tolerates high summer temperature and is good for hay. Crested Dog's-tail grows in many open areas throughout the woods.

Lesser Burdock *Arctium minus*

I think of this as a woodland ride plant, but there are none in the woods. Instead it appeared a few years ago on a solid clay bank in the Barn area and is spreading rapidly.

Marsh Marigold *Caltha palustris*

Despite the fact that we have perfect habitats for this lovely spring flower, none were present when I came here. I could not resist introducing a few where the spring that supplies us with water constantly overflows. Marsh Marigolds are now thriving, and some have grown into massive clumps. Note also the Watercress alongside, which was already established. Opposite-leaved Golden Saxifrage, Water Mint and Creeping Buttercup all jostle for space and light in this area, although I tend to thin out the latter whenever I pick the Watercress.

Water-starwort *Callitriche stagnalis*

I used to confuse this with Duckweed. Both appear in similar places (mostly in Pope's Wood) where water is more or less permanent. It is mat forming and prettily lights up shady woods with bright yellow patches when most other plants have turned dark green or even brown.

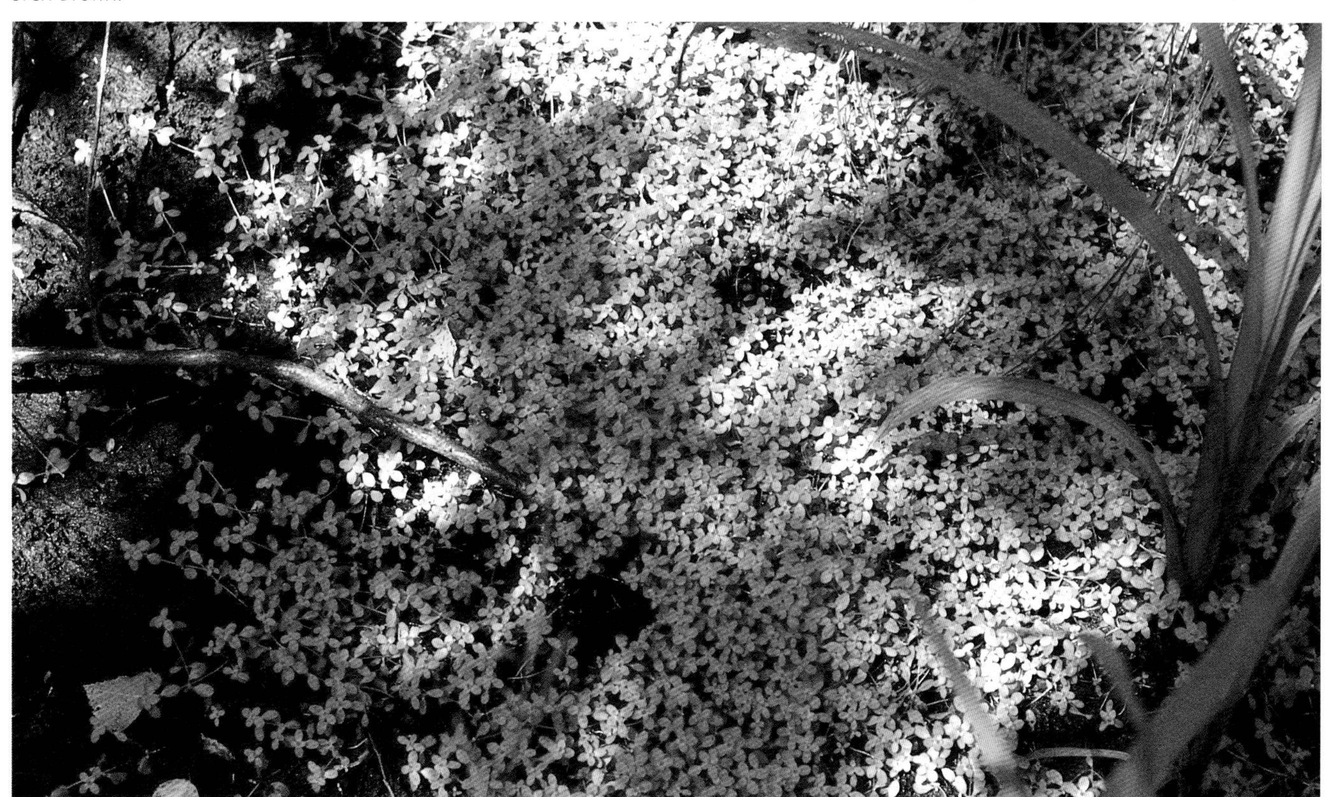

Nettled-leaved Bellflower *Campanula trachellum*

The flowers of all campanulas are lovely, as this picture illustrates. As I related in the text, at one time none grew here but I tried introducing it. This plant managed to flower, and I wonder whether its seeds will ever germinate. I would never persist in trying to establish a plant in the wild if it grew looking weak and sickly, but the two that survived appeared strong and healthy.

Pendulous Sedge *Carex pendula*

This is one of the largest and most graceful of the many sedges that grow here. These seem to grow anywhere there is bad drainage, although most are in Pope's Wood where in places they are becoming quite dominant.

Pignut *Conopodium majus*

I can remember as a child we used to pull up Pignuts and consume them with relish whilst out on school walks. The root is really very tasty. Today many would not only frown on this habit (fancy digging up something and eating it without first washing it!) but it would also be illegal. Pignut is an ancient woodland indicator and a member of the Umbilliferae *family, many species of which adorn the countryside in abundance with their white, lace-like flowers, notably Hogweed and Cow Parsley. The Pignut is much smaller and less abundant, although it is widespread here on most of the ancient woodland sites.*

Hard Fern *Blechnum spicant*

Many ferns are widespread, but this one confines itself to the banks of the streams in Badgers' Wood and Pope's Wood. It was gratifying to see new ones emerging on a stream bank in the latter that was once covered with Laurel.

Top right

Opposite-leaved Golden Saxifrage *Chrysosplenum oppositifolium*

What a grand name for a tiny plant whose flowers are only about 3 mm across! It forms large green/yellow carpets in any wet place. It is another ancient woodland indicator growing not only in Pope's Wood, but also in Streake's Wood and New Wood. Is this another sign of these two new woods evolving into traditional ancient ones?

Corn Marigold *Chrysanthemum segetum*

This picture really belongs in the butterfly chapter, but I could not resist putting it here. The flower is in our 'cornfield' (see page 107) and the Gatekeeper butterfly is really enjoying himself. Look at his stretched out legs and long proboscis delving into the brilliant gold of the flower – and no, the picture was not tampered with on the computer!

Enchanter's Nightshade *Circaea lutetiana*

These grow in all the woods and are one of my favourites. A single plant is unspectacular, but viewed en masse *– they usually grow in large ground-covering colonies – they are truly enchanting. If you go into the woods at dusk, or even in the dark with a torch, their white spires light up like a mass of tiny candles. It is a true woodland plant that evokes haunting and magic. If they got up and walked, it would not be difficult to imagine a crowd of sprites on the move.*

Wild Carrot *Daucus carota*

This plant only grows in the Barn area and occasionally in Shaw Wood. It is small and dainty and easily distinguished from the many other Umbellifers by its little red centre.

Marsh Thistle *Cirsium palustre*

Just look at all those spikes! They are in fact far less prickly than those of its 'nuisance' cousin the Creeping Thistle which is one of the *enfants terribles* in the garden. The tall Marsh Thistle, height 150cm or more, grows just about everywhere on our clay soil providing it can get sufficient light. Insects are attracted to it and it provides a feast of seeds for the finches.

Common Spotted Orchid *Dactylorhiza fuchsii*

No orchids are very common these days, but this one is becoming truly common in the woods. I think this is due to the fact that we keep a modicum of control over the invasive shade-creating plants like brambles and bracken. Unlike the Early Purple Orchid (see illustration on page 98), these flowers are extraordinarily tidy and uniform, creating an amazing mosaic-like pattern. Each floret is only a few millimetres across. They have also produced a self-generated meadow on the lawn (see illustration on page 22).

Wavy-hair Grass *Deschampsia flexuosa*

I feel this picture does not really do justice to the beautiful long seed stalks that glisten and wave in the sunshine. Its leaves are very fine and it grows up to 100 cm tall wherever the soil remains damp most of the year. The buds in the picture are sleek and silky and later turn into a spangled mass of tiny flowers like sparkling pinheads. I have noticed some being deliberately grown in cultivated flower beds, but I hope horticulturists will not change the colour or size of this beautiful grass.

Foxglove *Digitalis purpurea* (see also illustrations on pages 35 and 54)

It is a shame this magnificent plant is a biennial. It will not germinate in dense vegetation or without sufficient light. Some clearing and cutting has to take place (annually along the rides) so there is always the danger of destroying young plants. Each flower has a galaxy of spots to attract insects, and hairs to ensure pollen is rubbed on or off. The glycocides the plant contains are still used in modern and homeopathic medicines, mainly for heart ailments. The name comes from the Anglo-Saxon "foxes glova". I like the legend that bad fairies (one could call them good these days) gave the flower to the Fox to put on his feet so his prey would not hear him coming!

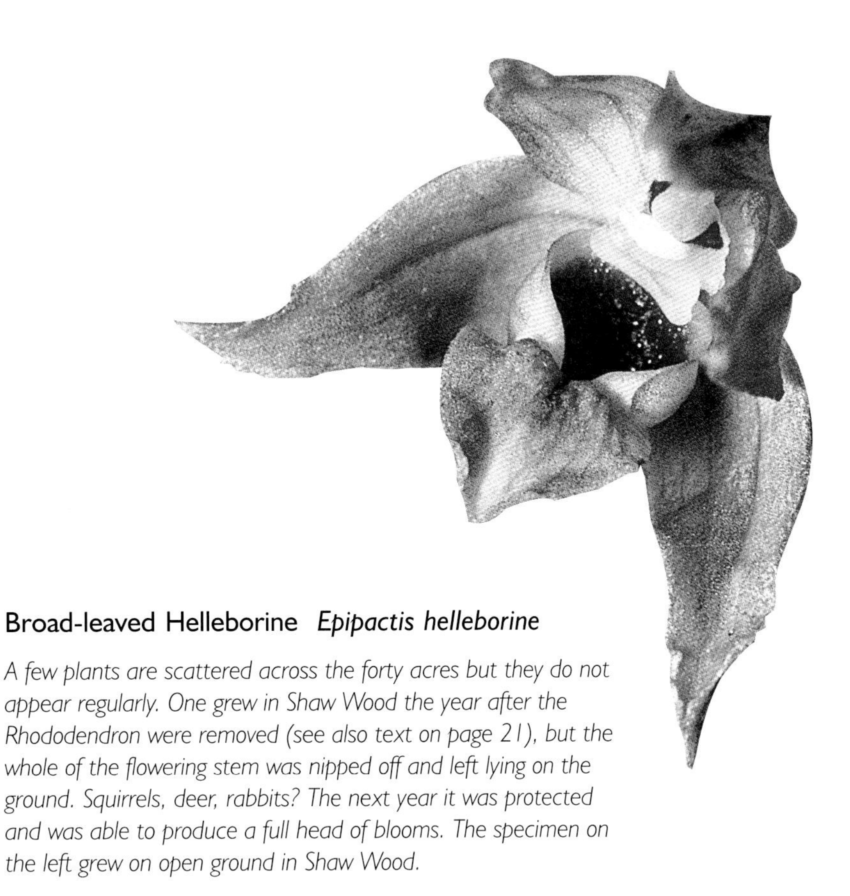

Broad-leaved Helleborine *Epipactis helleborine*

A few plants are scattered across the forty acres but they do not appear regularly. One grew in Shaw Wood the year after the Rhododendron were removed (see also text on page 21), but the whole of the flowering stem was nipped off and left lying on the ground. Squirrels, deer, rabbits? The next year it was protected and was able to produce a full head of blooms. The specimen on the left grew on open ground in Shaw Wood.

Dog's-tooth Violet *Erythronium dens-cani*

This is not a native, but I introduced a handful into Little Wood. Only one appeared and it comes up annually in an increasingly charming clump. It is an ideal woodland plant and certainly not invasive, so I had no qualms about introducing it.

Teasel *Dipsacus fullonum*

I have related the story about introducing Teasels on page 81. Unfortunately they look rather drab because 2003 was a particularly hot and dry summer – in fact, they were beginning to look stressed by the time they flowered. I wonder how they will fare in years to come.

Hemp Agrimony *Eupatorium cannabinum*

Hemp Agrimony grows tall and forms thick clumps. It can also become invasive – the whole chicken run is a forest of them unless I cut them as the chickens only dust-bathe in its shade! I try to leave them unless they are ousting particularly special plants, since all kinds of insects are attracted to them. I have seen bees, butterflies, hover-flies and countless others congregating on the flower heads as if at an insect AGM. The same is true of the tall Hogweed, *see page 94.*

Common Eyebright *Euphrasia nemorosa*

Apparently there are some twenty-five species of Eyebright as well as some hybrids growing in the UK. I think this is the only one growing here. They emerge in grass in midsummer as fair sized clusters, but each flower is only 2–3 mm across. It is delightful, but so tiny that it can easily be missed.

Right

Herb Robert *Geranium robertianum*

Opposite page

Bloody Cranesbill *G. sanguineum*

The delicately-leaved Herb Robert can be found everywhere which, unfortunately, cannot be said of Bloody Cranesbill whose magenta flowers (not really bloody) make it one of the brightest of wild plants. As it prefers chalky soils, it is hardly surprising that Bloody Cranesbill is sparse here – undoubtedly a garden escape. There are 800 species of Geranium in the world (not to be confused with the garden Pelargoniums) and most appear to be widely used in herbal medicine as a gentle astringent, a diuretic or for a host of other uses. The Germans have a good word for them, Augentrost*, meaning literally 'eye-comforter'. They were (and are?) also used in magic to promote peace and harmony.*

Ground Ivy *Glechoma hederacea*

This is another common plant of the labiate family worth a second look. The tiny flowers have gaping jaws with deep purple spots begging insects to come and visit. Note also the soft hairs which ensure that pollen will be rubbed off on different plants. It grows everywhere, forming ground-covering mats, unless with other vegetation when it can grow upright and tall.

Greater Bird's-foot-trefoil *Lotus uliginosus*

These are very popular with butterflies, especially the skippers and blues which, unlike most others, are hardly ever seen on the Buddleia, Stonecrop, Hemp Agrimony and others. They prefer this plant and White Clover which provide food for their larvae. These plants straggle everywhere in sunny places and in bud are prettily tinged red. The ubiquitous and much taller Rosebay Willowherb is threatening the trefoil, vetch, clover and other plant habitats. More cutting and pulling up at the roots is required!

Wild Gladiolus *Gladiolus illyricum ssp byzantinus*

*The native Wild Gladiolus (*Gladiolus illyricum*) supposedly grows only in the New Forest. Experts tell me this individual, which looks extremely like the wild variety, must be a garden escapee. It does grow on the edge of Shaw Wood, close to the garden, but it would have been nice to think it might have been a rare escapee – from the New Forest! Unfortunately, it only appears every four years or so, so no expert has actually seen it.*

Hogweed *Heracleum sphondylium*

Hogweed is a large plant that gets stronger every year until it is almost impossible to dig out. This can be very annoying in the garden unless it is caught young. It grows in my garden and everywhere else, but I tolerate it as it provides food and nectar for almost everyone in the insect world. See how the bees, hover-flies and others are clustered on it in the picture. Butterflies are also partial to Hogweed.

Above

Slender St. John's-wort *Hypericum pulchrum*

A number of St. John's-wort grow mainly in the south-facing glade in Shaw Wood. This glade is being taken over by the Rosebay Willowherb and virtually no Slender St. John's-wort appeared in 2003 so some extensive weeding will have to take place!

Below

Soft Rush *Juncus effusus*

The woods contain four species of rush. All of them grow in Pope's Wood but are widespread elsewhere also. They are present in either woods or rides wherever the soil is badly drained, or there is evidence of a spring. In very dry summers, like 2003, the presence of some sort of spring becomes very apparent: grasses and other plants turn brown in most areas but, in places, large stretches remain bright green no matter how dry the surrounding ground.

Left

Yellow Iris *Iris pseudacorus*

There was just one cluster of Yellow Iris in the shallow edge of Shaw Wood pond, but I have now increased this to many by planting some of the rhizones round the pond. It appears to be the only water plant robust enough to withstand the ducks on the pond. They have pulled up or eaten everything else I have planted.

Red Dead-nettle *Lamium purpureum*

I find the flowers of this plant quite captivating. Their domed red 'cyclists' helmets' really stand out, but look carefully at the forest of minute hairs: like most of the labiate family it has a protruding tongue (this one is spotted to make it even more visible) – the perfect alighting platform for insects.

Common Twayblade *Listera ovata*

I find that the term 'common' in the nomenclature of plants can be misleading since many of them are no longer common – certainly not orchids. I imagine the names go back a long way to when the countryside was very different. In this case I suppose it was to distinguish it from the Lesser Twayblade (L. cordata). 'Large Twayblade' might now be a better name. The name Twayblade comes from the two large leaves (below) from which the tall flowering stem (sometimes 50cm or more tall) emerges. The yellowish-green flowers are insignificant, but on closer examination they seem to take on a ghostly humanoid look. I have only found one small colony which grows on the border of Pope's Wood and Streake's Wood, and these I have protected from marauding grazers.

Wood Club-rush *Scirpus sylvaticus*

We have quite a number of rushes of which this is by far the largest. It is also an ancient woodland indicator. It grows up to 80cm tall and has very wide leaves. Although there are a few elsewhere, most of them grow in Pope's Wood where in one open patch there is a huge colony. They also grow in and around the small pond in this wood (see illustration page 38) where they compete healthily with Bur-reeds.

Ragged Robin *Lychnis flos-cuculi*

Ragged Robin is very erratic in appearance. This delightful plant is dishevelled and grows all over the place. It likes damp places, so I do not understand why it is not much more widespread here, nor why it is sometimes not seen for several years. A small dwarf version grows in Pope's Wood, but clumps like these are rare. They appeared one year on the ride down the middle of New Wood but have not been seen there since. Note the Marsh Bedstraw in the foreground. It is a great favourite with insects.

Purple Loosestrife *Lythrum salicaria*

I am at a loss as to why these lovely tall purple spikes only appear spasmodically by the pond in Shaw Wood. It seems to be a perfect habitat unless this particular area has become too shaded? I originally thought none were present here so I tried growing some elsewhere from seed. They flowered for a few years, but have now disappeared. Friends in Canada tell me the plant is causing concern there, becoming unduly invasive at the expense of local native plants. Attempts at eradication are proving difficult. This is yet another example of why wild plants should not be transposed from one country to another. Here, apart from the fact that its habitat is diminishing, something obviously controls them, whereas in a different climate they have gone mad.

Below

Early Purple Orchid *Orchis mascula*

What a contrast this orchid is to the Common Spotted Orchid (see page 89). Instead of the orderly pattern of the latter, this one has a helter-skelter of individual flowers looking in all directions. Yet viewed collectively (see page 25) they produce an orderly mass of lovely purple spikes. Tropical orchids have many devious ways of ensuring pollination, but most in Britain have these gaping mouths with large lips inviting insects to come and look inside.

Musk Mallow *Malva moschata*

Musk Mallows in the woods are very erratic both in appearance and quality. They are perennials but do not necessarily come up in the same place every year. I originally found them in Little Wood and Owl's Wood (all pink flowers) fighting their way through nettles and other vegetation. Some years later, this beautiful clump suddenly appeared at the edge of Owl's Wood, and many more are now in the garden where they doubtless find life much easier and where they are very welcome.

Wood Melick ***Melica uniflora***

This is one of the prettiest of the many grasses that inhabit the woods. It is small so can easily be overlooked unless it grows in large communities as it does in Shaw Wood. I found it impossible to photograph them in situ *and do them real justice. These were therefore photographed in my studio. Wood Mellick is another indicator of ancient woodland.*

Field Poppy or Corn Poppy *Papaver rhoeas*

Poppies, especially wild ones, must be some of the glories of nature the world over – their delicate, translucent paper-like petals and colour make them a wondrous sight, especially growing en masse. *They can only germinate where ground has been disturbed or is tilled every year. They and the lovely blue Cornflower no longer adorn our cornfields in the numbers that they used to. They have been all but eradicated from agricultural land in Britain by chemicals. It is a testament to their persistence that they still spring up, sometimes even along the banks of newly-made roads, only to disappear the following year because no one has disturbed the soil. They grow here in sunny sites wherever some soil disturbance has taken place and, of course, in our newly created 'cornfield'. These poppies are grown commercially, especially in Central Europe where they are much used in baking. In Austria people make a strudel, filled not with apples, but a delicious poppy seed stuffing.*

Wood Sorrel *Oxalis acetosella*

Surely this must be one of the most endearing of all woodland plants with its fragile, delicately veined flower and clover-like leaves. It grows in many of our woods but flourishes best in Pope's Wood. There it can be found in large clusters in moss or leafmould, and even on trees. This one is growing in the fork of two large Ash trunks. The plant has inspired many to write in its praise, and it has also acquired many evocative names: Fairy Bells, Hallelujah, Cuckowes Meat and Three-leaved Grass to name a few.

Solomon's-seal *Polygonatum multiflorum*

I originally found one clump in the orchard, but as it was continually eaten by the animals, I moved it to the edge of Shaw Wood and Pope's Wood. Although a native plant, it may not have grown here naturally, but it seems odd that someone should have planted it in the orchard! It is a lovely plant and I am glad it thrives in its new home.

Common Polypody
Polypodium vulgare

As far as I know, these only grow on the west facing bank in Shaw Wood. It is another species that could do without the prefix of 'common'. Although not endangered, it is certainly not that common. It is an ancient woodland indicator.

Lesser Spearwort *Ranunculus flammula*

I welcome this member of the buttercup family. We have many of them in wet places in Butler's Wood and Badgers' Wood, and they are not as invasive as some other members of the family. The Meadow Buttercup is lovely, as is the Marsh Marigold and the Lesser Celandine, but the Creeping Buttercup can be a real nuisance, especially if you a trying to grow grass for grazing (pretty sparse here) as animals will not touch them! However, we have found by mowing the grazing areas, the number of Creeping Buttercup has declined.

White Campion *Silene latifolium*

We have many Red Campion throughout the forty acres, but this white variety grows, as far as I am aware, only in one spot at the edge of Owl's Wood.

Sanicle *Sanicula europaea*

An ancient woodland indicator that grows in Pope's Wood and Shaw Wood. It was widely used for medicinal purposes. Its name comes from the Latin sano, *I cure.*

Goldenrod *Solidago virgaurea*

*This is not the Goldenrod of gardens (*Solidago canadensis*) which is often found naturalised, but a native ancient woodland plant. It flowers in late summer/autumn. Here it is found mainly in Shaw Wood where it is spreading and will doubtless soon appear in the other woods.*

Self-heal *Prunella vulgaris*

Here is another member of the labiate family with gaping jaws and fine detail inside. It seems specially made for bees to insert their long proboscises. It is common here and everywhere.

Hedge Woundwort *Stachys sylvatica*

I think the intricacy of these flowers competes with that of many orchids. Here they are greatly enlarged as they are only 2–3 mm across, but nevertheless it demonstrates what can be discovered if they are looked at closely. Don't forget to take a magnifying glass if going on wild flower forays, because it would be a pity to miss the minutiae of this plant or, for that matter, many others. The whole plant is sometimes hard to spot growing among other flowers and grasses where it seems insignificant, but what a revelation close up!

Below

Betony *Stachys officinalis (formerly Betonica officinalis)*

There are a number of colonies appearing in Shaw Wood and also in the many places in the garden where the grass is cut only once a year. So far, I have not noticed them in any of the other woods. The flower-head is truly unruly, like the Early Purple Orchid, and reminds me more of some new, wild hairstyle, rather than an ancient woodland indicator!

Opposite top

Smooth Sow Thistle *Sonchus oleraceus*

I can imagine some saying this picture does not really belong here, but it does illustrate individual diversity. Although widespread, this one grew in the rich soil of the vegetable garden. I found it and another tinged with slightly different colours and still wet with dew one morning and could not resist photographing it. It is an example of what can happen to a wild plant if suddenly given unlimited space and a larder so full of nutrients that it does not know what to do! In the real wild, it is insignificant and not that atttractive.

Opposite bottom

Wood Sage *Teucrium scorodonia*

*This is common in most of the woods but grows particularly well in Owl's Wood where there are many dry places. The leaves resemble those of the culinary Sage (*Salvia officinalis*), which have grown in Britain for many centuries but which originate from Southern Europe. I imagine, therefore, that it was the latter that was used medicinally and in cooking rather than the Wood Sage. However, it is said that an infusion of Speedwell, Sage and Betony makes an excellent breakfast drink.*

Common Nettle *Urtica dioica*

I mentioned this plant in the introduction (see page 15) but did not stress what an important role it plays in biodiversity. Animals may not graze it, but it is the food plant of countless moths, butterflies and other insects. The tiny hairs, sometimes as big as thorns, squirt out an irritant immediately it is touched. This is why animals never eat it (unless first cut and left, after which they relish it – it has many nutrients) and why humans feel as if they have been stung, or even develop a rash. The way to avoid this is to grab the plant so quickly it has no time to release the irritant, but on the whole it is probably wiser to wear strong gloves. It is specially partial to nitrogen-rich soil, so it can also become a problem if it grows dense and widespread, ousting other smaller plants that may prefer a less rich diet. As we only have a few chickens and two (sometimes three) Shetland ponies, it causes minor problems.

Left

Marsh Violet *Viola palustris*

I am afraid this picture does not do justice to this little violet, but I have included it because, unlike most of the other Viola species, it is becoming scarce, largely due to shrinking habitat. It likes shady, wet, boggy soil, of which we have plenty in Pope's Wood and where we are lucky to have several large colonies.

Our 'cornfield'

I am concluding this plant section with a picture of a 'meadow' we have created of cornfield flowers. I have included some of the plants that grow in it in the previous pages. They include Corncockle, Corn Marigold, Poppy, Corn Chamomile and the very beautiful Cornflower. The last two are not very visible since they are much shorter and are somewhat overshadowed by the others. The field is only in its second year.

Fungi, Mosses and Lichens

I wonder how many people are aware of the paramount importance of fungi, not just in our daily lives, but also in all of nature? Most of us are familiar with the fungi that grow in our woods and fields and that either send us into ecstasies of gourmet delight, or cause violent indigestion, hallucinations and sometimes, inadvertently, even death. We are also quite vitriolic when fungi undermine the structure of our buildings (dry rot, for example, caused by *Serpula lacrymans*) or when they produce obnoxious, dripping exuberances in our rooms. Our fury is vented on the fungi, often forgetting it is usually ourselves who are to blame for not keeping rooms dry and ventilated. Many virulent fungi attack trees and plants and often destroy them.

However, the all-important fungi are the invisible mycorrhiza, without which most of the natural world would not be as we know it. They are ubiquitous from the Antarctic to tropical rain forests but exist as invisible hyphae both in the soil and roots and stems of plants. More than ninety per cent of the world's plant species rely on them for nutrients such as nitrogen, phosphorus, trace elements, minerals and water, which they take up through their roots. Orchids and other plants cannot even germinate in their absence. In return, the plants' roots supply the mycorrhiza with the carbohydrates essential to their well-being, thus forming a truly symbiotic relationship. They are extremely sensitive to all chemicals, which will destroy them and other soil organisms. Experiments show that although you may be able to grow a woodland (or any other type of ecosystem) on degenerated chemically polluted arable farmland, the chance of producing a diverse understorey are remote – it could take centuries.*

I mentioned in the introduction how fortunate it was that my forty acres consisted largely of ancient woodland sites. The only ones that were once arable land (Streake's Wood, New Wood and Morgan's Strip) probably avoided the explosion of chemicals in the mid-twentieth century, so they must have a greater chance of becoming a truly biodiverse ecosystem in a relatively short time. The days when farmers had fewer mechanical aids and had to rely on farmyard muck and humus for their crops cannot but have made an enormous difference. I am sure that, although some plants such as Primrose, Bugle, Red Campion and others do not depend on the mycorrhiza, it is due to the presence of the last that the woods are so diverse in plant life.

The 'visible fungi' (the subject of the survey) are not, like some of the other organisms in this section, in any way symbiotic. On the contrary, their existence depends on rotting vegetation and dung, and some are even parasitic. My dealings with these fungi go back a very long way, but regrettably they were confined to the edible kind and only the delicious ones at that. The others, or ones that I did not know, were largely ignored. My knowledge was and is, therefore, scant indeed. Over the years, whenever an interesting species presented itself, I would photograph it and attempt identification. This posed some problems to Peter Russell who carried out the survey. Some, like the Coral or Bearded Tooth Fungus are distinctive, but others were and still are a puzzle. Peter came on many days, but only in 2003, so the survey is a mixture of a few species identified from my pictures and the rest found by Peter during that year. The fact that he was able to produce 157 species is something of a miracle, especially as the summer of 2003 was particularly dry and hot, and even early autumn brought no rain. Apart from that, many fungi appear only infrequently and when it suits them. I know that over a period of several years, the number of species would run into many hundreds. Despite the 'lean' 2003, one fact did emerge: we appear to have several uncommon species present. Peter has given me the actual number of some as have been officially recorded by the British Mycological Society (BMS). They appear in brackets in the comments column of the survey.

As far as mosses and liverworts are concerned, I have always admired their velvety greenness and delicate feathery 'leaves' and been astonished at their virtually instant revival after prolonged dry periods. Mosses rely entirely on moisture from which they extract nutrients, which is why they are so abundant in Pope's Wood and also elsewhere due to the water-retentive clay on which all the woods grow.

If my knowledge of fungi and mosses was scant, it was virtually non-existent as far as lichens were concerned. The structure of lichens is unique – a mutualistic symbiotic relationship of two other organisms, a fungus and an alga. Lichens grow anywhere, including on walls and fences, and are able literally to eat into stones and rocks. They can survive extreme cold and heat and can remain dormant for many years. Some can endure temperatures of over 50°C, whilst others form an important component in Arctic ecosystems.

Until recently lichens have been somewhat neglected, which is probably why they have not so far acquired common names. It seems they are behind the times – on the other hand, bryophytes have recently all acquired very fancy, romantic sounding common names!

I was aware that many lichen species require clean air and supposed that they were declining, as even here we cannot escape pollution. I am told, however, some lichens have reappeared in London! At least this is a sign of cleaner air, although some species are more tolerant to pollution than others. In fact, *Lecanora conizaeoides* actually seems to like it! Nevertheless, lichens are extremely sensitive to sulphur dioxide – it destroys the chlorophyll in the algal partners and they themselves are very efficient absorbers of anything in the air.

Shady, dense woodlands may not be an ideal lichen habitat, so it was a real surprise when Jacqui Middleton meticulously prised out no less than sixty-three species.

Malcolm McFarlane and Jacqui and Bruce Middleton made the survey of the mosses, liverworts and lichens over many days in 2003.

* I am indebted for much of the information in this paragraph to Jeremy Merryweather, *British Wildlife* Vol. 13, no 3.

Fungi

Scientific names are from the current BMS list

Common names are from the current list of 'Recommended English Names for Fungi' by the BMS, English Nature, Plantlife and others

Scientific Name	Common Name	Comments
Agaricus campestris	Field Mushroom	Frequent in orchard
Amanita citrina	False Deathcap	Widespread and frequent
Amanita excelsa	Grey Spotted Amanita	Widespread and frequent
Amanita fulva	Tawny Grisette	Found in Butler's Wood and Owl's Wood
Amanita muscari	Fly Agaric	Widespread and frequent
Amanita rubescens	Blusher	Widespread. See illustration p112.
Amanita spissa	Grey Spotted Amanita	Infrequent. See illustration p112.
Armillaria gallica	Bulbous Honey Fungus	Widespread and frequent
Armillaria mellea	Honey Fungus	Widespread and frequent
Bjerkandera adusta	Smoky Bracket	Mainly in Pope's Wood
Bolbitius vitellinus	Yellow Fieldcap	In long grass around lawn
Boletus appendiculatus	Oak Bolete	Frequent in Shaw Wood (156 records)
Boletus edulis	Penny Bun/Cep	Frequent but not every year
Boletus queletii	Deceiving Bolete	Frequent mainly in Shaw and Butler's Wood
Boletus radicans	Rooting Bolete	Infrequent. See illustration pp112-113.
Boletus rubellus	Ruby Bolete	Infrequent. See illustration pp112-113.
Calocera viscosa	Yellow Staghorn	Mainly Badger's Wood
Calvatia excipuliformis	–	Infrequent in Shaw Wood
Calvatia gigantea	Giant Puffball	Sometimes in great numbers in orchard
Cantharellus cibarius	Chanterelle	Frequent and normally in Shaw Wood
Chlorociboria aeruginascens	Green Elfcup	See illustration p115
Chondrostereum purpureum	Silverleaf Fungus	Occasional in Pope's Wood
Clavariadelphus fistulosus	–	Single specimen in Pope's Wood
Clavulina cristata	–	On ground in Pope's Wood
Clitocybe infundibuliformis	–	Mainly in Pope's Wood
Collybia butyracea	Butter Cup	Common in Shaw Wood
Collybia dryophila	Russet Toughshank	Ocassional in Shaw Wood
Coprinus comatus	Shaggy Inkcap/Lawyer's Wig	Frequent especially in Little Wood and Owl's Wood
Coprinus micaceus	Glistening Inkcap	Frequent especially in Little Wood and Owl's Wood
Coprinus niveus	Snowy Inkcap	Dung heaps
Cortinarius pseudosalor	–	Infrequent in Pope's Wood
Cortinarius sanguineus	Bloodred Webcap	Occasional in Pope's Wood
Cortinarius triumphans	Birch Webcap	Infrequent in Pope's wood

Scientific name	Common name	Comments
Crepidotus mollis	Pealing Oysterling	Common
Crepidotus variabilis	Variable Oysterling	Common
Daedaleopsis confragosa	Blushing Bracket	Frequent in Pope's Wood
Daldinia concentrica	King Alfred's Cakes	Pope's Wood
Exidia thuretiana	White Brain	Infrequent – Owl's Wood
Fistulina hepatica	Beefsteak Fungus	Fairly widespread
Flammulina velutipes	Velvet Shank	Widespread
Ganoderma australe	Southern Bracket	Widespread
Gymnopilus junonius	Spectacular Rustgill	See illustration p117
Hebeloma sacchariolens	Sweet Poisonpie	Frequent in Pope's Wood
Hericium erinaceum	Coral Fungus/Bearded Tooth	See illustration p116 (251 records)
Heterobasidion annosum	Root Rot	Morgan's Strip
Hydnum repandum	Wood Hedgehog	See illustration p114
Hygrophoropsis aurantiaca	False Chanterelle	Frequent, especially in Badgers' Wood
Hymenochaete rubiginosa	Oak Curtain Crust	Occasional in Pope's Wood (81 records)
Hypholoma fasciculare	Sulphur Tuft	Widespread – see illustration p115
Hypholoma sublateritium	–	Widespread
Hypoxylon fragiforme	Beech Woodwart	Mainly Shaw Wood
Hypoxylon multiforme	Birch Woodwart	Pope's Wood
Inonotus radiatus	Alder Bracket	Infrequent – Pope's Wood
Kuehneromyces mutabilis	Sheathed Woodtuft	Mainly in Pope's Wood
Laccaria amethystina	Amethyst Deceiver	Frequent in Shaw Wood and Butler's Wood
Laccaria laccata	Deceiver	Widespread (29 records)
Lactarius azonites	–	Infrequent – Shaw Wood (81 records)
Lactarius circellatus	–	Mainly under Hornbeam
Lactarius deterrimus	False Saffron Milkcap	Mainly Badgers' Wood (98 records)
Lactarius glyciosmus	Coconut Milkcap	Widespread under Birch
Lactarius piperatus	Pepper Milkcap	Frequent
Lactarius quieticolor	–	Frequent in Pope's Wood
Lactarius quietus	Oakbug Milkcap	Occasional in Pope's Wood
Lactarius tabidus	Birch Milkcap	Frequent in Pope's Wood
Lactarius torminosus	Woolly Milkcap	Infrequent in Pope's wood
Lactarius turpis	Ugly Milkcap	Frequent in Pope's Wood
Lactarius zonarius	–	Infrequent in Shaw Wood (29 records)
Laetiporus sulphureus	Chicken of the Woods	Fairly frequent
Leccinum carpini	–	In Shaw Wood under Hornbeam (64 records)
Leccinum crocipodium	Yellow-cracking/Saffron Bolete	See illustration p114 (98 records)
Leccinum pulchrum	–	Infrequent in Butler Wood
Leccinum scabrum	Brown Birch Bolete	Widespread
Lenzites betulina	Birch Mazegill	Frequent in Pope's Wood
Lepista inversa	–	Occasional in Pope's Wood
Lepista nuda	Wood Blewit	Sometimes in large numbers
Lepista saeva	Field Blewit	Most years near house
Lycoperdon perlatum	Common Puffball	Occasional in Pope's Wood
Macrolepiota konradii	–	Occasional in Shaw Wood
Macrolepiota procera	Parasol	Frequent, and sometimes in great numbers
Macrolepiota rhacodes	Shaggy Parasol	Frequent most years
Marasmiellus ramealis	Twig Parachute	Widespread
Megacollybia platyphylla	Whitelaced Shank	Frequent in Pope's Wood
Mycena galericulata	Common Bonnet	Frequent on old wood stumps
Mycena galopus	Milking Bonnet	Common in Pope's Wood
Mycena inclinata	Clustered Bonnet	Common in Pope's Wood
Mycena maculata	–	Normally on dead Beech
Mycena polygramma	Grooved Bonnet	Frequent on old twigs
Mycena viscosa	–	Common in Pope's Wood
Mycena vitilis	Snapping Bonnet	Frequent on branches and broadleaf bark
Naucoria escharoides	Ochre Aldercap	Frequent, especially Pope's Wood
Nectria cinnabarina	Coral Spot	Frequent on dead branches
Oxyporus populinus	Poplar Bracket	Frequent on old broadleaf wood
Panaeolus fimicola	Turf Mottlegill	Grey fungi frequent in grass
Panaeolus semiovatus	Egghead Mottlegill	Frequent in grass
Paxillus involutus	Brown Rollrim	Fairly frequent
Peziza badia	Bay Cup	Along ride in Shaw Wood (3 records)

Scientific Name	Common Name	Comments
Phaeolus schweinitzii	Dyer's Mazegill	Frequent in Morgan's Strip
Phallus impudicus	Stinkhorn	Found in all the woods (62 records)
Phlebia radiata	Wrinkled Crust	Frequent on dead wood
Phlebia tremellosa	Jelly Rot	Frequent on dead wood
Pholiota flammans	Flaming Scalycup	Infrequent on conifer stumps
Piptoporus betulinus	Birch Polypore	Very frequent everywhere
Pleurotus ostreatus	Oyster Mushroom	Fairly frequent on dead wood
Pleurotus pulmonarius	Pale Oyster	On logs and stumps of broadleaves (76 records)
Pluteus cervinus	Deer Shield	Mainly in Pope's Wood
Pluteus salicinus	Willow Shield	Frequent in Pope's Wood
Pluteus umbrosus	Velvet Shield	Old trunks mainly in Pope's Wood (2375 records) (common)
Polyporus badius	–	Infrequent in Pope's Wood
Polyporus ciliatus	Fringed Polypore	Infrequent in Pope's Wood
Postia stiptica	Bitter Bracket	Mainly on dead pine
Psathyrella candolleana	Pale Brittlestem	Rotting branches in Pope's Wood
Psathyrella microrhiza	Rootlet Brittlestem	In Shaw Wood
Rickenella fibula	Orange Mosscap	Occasional in Shaw Wood
Russula amarissima	–	Single specimen in Shaw Wood (3 records)
Russula betularum	Birch Brittlegill	Frequent in Butler's Wood
Russula brunneoviolacea	–	Infrequent in Butler's Wood (15 records)
Russula cyanoxantha var. peltereaui	Charcoal Burner	Frequent in Pope's Wood
Russula emetica f. sylvestris	Sickener	One clump in Shaw Wood (80 records)
Russula heterophylla	Greasy Green Brittlegill	Infrequent in Shaw Wood
Russula ionochlora	Oilslick Brittlegill	One specimen in Shaw Wood
Russula laurocerasi	–	Frequent in Shaw Wood and Butler's Wood
Russula luteotacta	–	On side of drainage ditch in Shaw Wood (76 records)
Russula mairei	–	In Shaw Wood
Russula nitida	Purple Swamp Brittlegill	Mainly in and around Shaw Wood
Russula ochroleuca	Ochre Brittlegill	Frequent in Pope's Wood (2735 records)
Russula pectinata	–	One specimen in Shaw Wood
Russula pelargonia	–	Mainly in and around Shaw Wood
Russula pseudointegra	Scarlet Brittlegill	Frequent in Shaw Wood and Butler's wood
Russula risigallina	Golden Brittlegill	One specimen in Shaw Wood
Russula romellii	–	One specimen in Shaw Wood
Russula rosea	Rosy Brittlegill	Infrequent in Shaw Wood
Russula sanguinea	Bloody Brittlegill	Occasional in Pope's Wood
Russula sororia	Sepia Brittlegill	Mainly in and around Shaw Wood
Russula subfoetens	–	Frequent in Shaw Wood and Butler's Wood
Russula urens	–	One specimen in Shaw Wood
Russula vesca	The Flirt	Mainly in and around Shaw Wood
Sarcoscypha coccinea	Scarlet Elfcup	See illustration p116
Schizophyllum commune	Common Porecrust	Single clump in Pope's Wood
Scleroderma citrinum	Common Earthball	Widespread and frequent
Skeletocutis nivea	Hazel Bracket	Widespread on branches of broadleaves
Sparassis crispa	Wood Cauliflower	Fairly frequent usually near conifers
Stereum hirsutum	Hairy Curtain Crust	Frequent on dead wood
Stropharia semiglobata	Dung Roundhead	Normally in compost
Suillus bovinus	Bovine Bolete	Frequent among the Scots Pine
Suillus luteus	Slippery Jack	Frequent in grass near house
Thelephora penicillata	–	On soil along ride in Shaw Wood
Trametes hirsuta	Hairy Bracket	Infrequent on tree trunks – found in Owl's Wood
Trametes versicolor	Turkeytail	Frequent on tree trunks – found in Owl's Wood
Tremella mesenterica	Yellow Brain	Infrequent on tree trunks – found in Owl's Wood
Tricholoma virgatum	Ashen Knight	Fairly widespread, normally near conifers
Tylopilus felleus	Bitter Bolete	Fairly frequent in all the woods
Xerocomus badius	Bay Bolete	Frequent in Shaw Wood
Xerocomus chrysenteron	Red Cracking Bolete	Frequent under broadleaves
Xerocomus subtomentosus	Suede Bolete	Occasional under broadleaves

Above
The Blusher *Amanita rubescens*

Left
Grey Spotted Amanita *Amanita spissa*

Nearly all the amanitas are poisonous unlike the boletes, (most of which are edible) and the Grey Spotted Amanita is no exception. The Blusher is one of the few edible amanitas and can be identified by the absence of the vulva, out of which the fungi appears to grow (seen clearly on the Grey Spotted), and by the fact that the flesh turns pink when cut or bruised. However, I have never trusted myself to eat it as, often, when all else seemed right, the flesh did not turn colour – maybe I was too impatient and did not wait long enough!

Opposite top
Rooting Bolete *Boletus radicans*

Opposite below
Ruby Bolete *Boletus rubellus*

The summer and early autumn of 2003, when the survey was in progress, was particularly dry and hot and not ideal for a proliferation of fungi. Nevertheless, Peter found quite a number and often remarked he was finding unexpected species. This Ruby Bolete was one of them but in that year it only made a scant appearance. It appears faithfully every year against a wall of the house, under an old specimen of a pendulous cotoneaster. Boletus radicans *appears sporadically on the ride between Shaw Wood and New Wood.*

Wood Hedgehog *Hydnum repandum*

This is a delicious fungus and, in my opinion, a close rival to the more famous Chanterelle. Raw, it tastes sharp and bitter but the bitterness disappears when cooked. Most years it grows in Shaw Wood, but its numbers are erratic. In 2002, two years after we had ousted all the Rhododendron and Laurel growing in Shaw Wood, it appeared in huge quantities where there had once been a large Laurel. Neither the Wood Hedgehog nor the Cow-wheat (see also page 21) appear to have been affected by the poison that Laurel produces (prussic acid), nor indeed was I, as I had the fungus on my menu for several weeks! The odd fact was that they did not appear in that spot in 2003 – maybe it was too dry.

Yellow-cracking or Saffron Bolete *Leccinum crocipodium*

As scientific techniques advance into the DNA and other properties of all flora, the classification and nomenclature is undergoing upheavals. Some time ago it was the turn of plants, and reference books and people are only just beginning to catch up with this on-going revolution. In the world of fungi the situation is even worse. Not only does one have to grapple with two, three or more different names for each species, but also it seems that the species groups are being broken up and this Leccinum *may be one of them. However, I am told the situation was always so, and I wonder how even mycologists can cope. This species appears regularly but not frequently in the woods.*

Green Elfcup *Chlorociboria aeruginascens*

Cabinet makers in Tunbridge Wells, not far from here, formerly used this greenish stained wood for furniture making and inlays. Dead pieces of wood stained this colour are often found, but the fruiting cups are much rarer. Peter brought in this piece of wood, and it still sits outside the house on damp shaded earth producing more fruit as others die off.

Sulphur Tuft *Hypholoma fasciculare*

This is indeed like fairyland! The fungi look just like a gathering of pixies debating their next move. What is more, the Sulphur Tufts are growing on the fringes of Pope's Wood (the one that I consider enchanted) on a dark hilly strip just inside Butler's Wood. The Sulphur Tuft is common here among Birch and Hornbeam and often appears in great numbers everywhere but rarely, I think, in such quantities.

Scarlet Elfcup *Sarcoscypha coccinea/austriaca*

I can find this species annually on dead wood in the stream separating Badgers' Wood from Owl's Wood. It is a pretty 'cup' fungus and said to be edible, but I have never tried it.

Wood Cauliflower *Sparassis crispa*

These fungi can indeed look like cauliflowers, and occasionally they appear in large quantities. They nearly always grow under the Scots Pines, of which we have many. They are also very good to eat although cleaning them can sometimes be a problem as their strange structure collects all kinds of debris.

Coral Fungus/Bearded Tooth *Hericium erinaceum*

This stunningly beautiful fungus first appeared around 1990 on a pile of large Beech trunks that were left in Shaw Wood after The Storm. The fungus is now rare and protected throughout the European Union countries, but I can remember collecting it many decades ago in Europe (when it was still abundant) and relishing it in many a meal. Unfortunately, it appeared only once again in 1991 and never since. I hope the mycelium, or spores, are still around and that it will appear again one day.

Spectacular Rustgill
Gymnopilus junonius

I have not seen this fungus very often, certainly not in such a dense cluster. This clump grew round the base of a dead Scots Pine. Although not poisonous, it tastes very bitter and is not edible.

Chicken of the Woods
Laetiporus sulphureus

You can often find this spectacular fungus growing on old tree stumps, or the trunks of trees that have been damaged in some way. It is a true parasite, rotting the wood within the trunk. This particular one has appeared annually on the same broadleaf trunk on the border of Owl's Wood. The name 'Chicken of the Woods' comes from the fact that the flesh texture is similar to chicken meat. It is considerably tougher than the modern factory-produced chicken meat, but good to eat nonetheless.

Mosses

Scientific name	Common name	Comments
Atrichum undulatum	Common Smoothcap	Widespread
Aulacomnium androgynum	Bud-headed Groove-moss	Mainly on rotting wood
Brachythecium rutabulum	Rough-stalked Feather-moss	Widespread
Bryum capillare	Capillary Thread-moss	Widespread
Calliergon cuspidata	Pointed Spear-moss	Widespread
Campylopus introflexus	Heath Star Moss	Alien: first appeared in the UK in Sussex, 1941
Ceratodon purpureus	Redshank	Mainly in Barn area
Cirriphyllum piliferum	Hair-pointed Feather-moss	Grows in grass
Cryphaea heteromalla	Lateral Cryphaea	Widespread on Elder
Dicranella heteromalla	Silky Forklet-moss	Widespread
Dicranoweisia cirrata	Common Pincushion	Widespread
Dicranum scoparium	Broom Fork-moss	Widespread
Eurhynchium praelongum	Common Feather-moss	See illustration p121
Eurhynchium striatum	Lesser Striated Feather-moss	Widespread
Fissidens bryoides	Lesser Pocket-moss	Widespread
Fissidens taxifolius	Common Pocket-moss	Widespread
Frullania dilatata	Dilated Scalewort	Liverwort: widespread
Homalothecium sericeum	Silky Wall Feather-moss	Walls and trees
Hookeria lucens	Shining Hookeria	See illustration p119
Hypnum andoi	Mamillate Plait-moss	Widespread, mainly on Beech
Hypnum cupressiforme	Cypress-leaved Plait-moss	Widespread
Hypnum jutlandicum	Heath Plait-moss	Mainly Badgers' Wood. See illustration p119.
Hypnum resupinatum	Supine Plait-moss	Widespread on tree trunks
Isothecium myosuroides	Slender Mouse-tail Moss	Widespread
Lophocolea bidentata	Bifid Crestwort	Liverwort; widespread on tree stumps
Lophocolea heterophylla	Variable-leaved Crestwort	Liverwort; widespread on rotting wood
Lunularia cruciata	Crescent-cup Liverwort	Liverwort; widespread on tree stumps
Marchantia polymorpha	Common Liverwort	Liverwort; see illustration p122
Metzgeria furcata	Forked Veilwort	Liverwort; widespread
Miccrolejeunea ulicina	Fairy Beads	Liverwort; widespread
Mnium hornum	Swan's-neck Thyme-moss	Widespread. See illustration p119.
Orthodontium lineare	Cape Thread-moss	Alien from South Africa
Orthotrichum affine	Wood Bristle-moss	Widespread on trees
Orthotrichum anomalum	Anomalous Bristle-moss	Rocks and walls
Pellia epiphylla	Overleaf Pellia	Liverwort: widespread on stream banks
Plagiochila asplenioides	Greater Featherwort	Liverwort
Plagiomnium undulatum	Hart's-tongue Thyme-moss	Widespread
Plagiothecium sp	Silk Moss	Widespread
Polytrichum commune	Common Haircap	Widespread
Polytrichum formosum	Bank Haircap	Fairly widespread. See illustration p120.
Polytrichum juniperinum	Juniper Haircap	Widespread
Pseudotaxiphyllum elegans	Elegant Silk-moss	On banks in Barn area
Radula complanata	Even Scalewort	In Barn area on Ash
Rhizomnium punctatum	Dotted Thyme-moss	Pope's Wood
Rhytidiadelphus squarrosus	Springy Turf-moss	Widespread
Riccia glauca	Glaucus Crystalwort	Liverwort
Scleropodium purum	Neat Feather-moss	Barn area
Sphagnum palustre	Blunt-leaved Bog-moss	Pope's Wood only. See illustration p121.
Sphagnum sp	Bog-moss	Pope's Wood only
Thuidium tamariscinum	Common Tamarisk-moss	See illustration p120
Tortula muralis	Wall Screw-moss	On sandstones
Ulota crispa	Crisped Pincushion	On tree trunks and branches

Swan's-neck Thyme Moss *Mnium hornum*

These are the 'flowers' or capsules (hugely enlarged) of one of the commonest mosses. Here they are widespread, often forming large, dense swards especially in Shaw Wood. Unlike most other plants, the buds are produced in autumn and die out in spring. The tree-like trunk in the background is, in fact, only a small Beech seedling.

Heath Plait-moss *Hypnum jutlandicum*

*In Shaw Wood there is a large hump-backed peninsula jutting out into the pond (part of it is shown on the opposite page). There are many mosses on this bank but the Heath Plait-moss dominates. Many of the oaks, hornbeams, beeches and field maples have Slender Mouse-tail Moss (*Isothecium mysuroides*) climbing from the base of the trunks. The picture shows Heath Plait-moss flowering in December. Winter is the main fruiting season for mosses and liverworts.*

Shining Hookeria *Hookeria lucens*

The discovery of this moss caused great excitement for Malcolm. It was not so much that it is uncommon in this part of Britain, but that there were large colonies of it in Pope's Wood. It is a peculiarly unphotogenic moss!

Bank Haircap *Polytrichum formosum*

This is an easy moss to identify, with its spiral spikes not seen in most other mosses. It is widespread here and quite easy to find where conditions are reasonably damp. It prefers acidic soil and can therefore often be found on heathlands. The soil in our woods is not particularly acidic: it is neutral or only slightly acidic.

Below

Common Tamarisk-moss *Thuidium tamariscinum*

An extremely pretty moss that is easy to identify: its 'leaves' are quite large compared to other mosses and very distinctive with their feathery, fern-like appearance. This moss is widespread here but only in abundance where the soil remains wet even during drought years.

Tree root

This large tree stump with some of the roots attached is all that remains of a large Douglas Fir in Owl's Wood that came down in The Storm. Although not a good picture, I am including it because it is a wonderful example of the biodiversity of nature on what is, after all, only an old root. As well as some grass and puny brambles, the following nine flora were found on it:

Moss: Atrichum undulatum
Alien Moss: Campylopus introflexus
Lichen: Cladonia conicrea
Lichen: Cladonia fimbriata
Moss: Dicranoweisia cirrata
Moss: Eurhynchium praelongum
Moss: Hypnum cupressiforme
Liverwort: Lophocolea heterophylla
Alien Moss: Orthodontium lineare

Bog-moss *Sphagnum palustre*

There are many species of Sphagnum *or bog-mosses growing in the UK. We have at least two growing here and maybe more. In Pope's Wood the bog-moss above grows over an area of several square metres. It grows alongside the 'meadow' of Wood Club-rush, with many mosses creeping in and out of this lovely plant. Bog-mosses are much used in garden centres for moss poles, hanging baskets and others – a practice which I hope will decline.*

Common Liverwort
Marchantia polymorpha

This is a thalloid liverwort (there is no differentiation between leaves and stems) that is commonly found in badly drained flowerpots and elsewhere. However, common or not, it is fascinating if viewed with a magnifying glass. It is dioecious: it has separate male and female parts. In the picture, which is greatly enlarged, the cup-like structures in the foreground are the males and the little 'umbrellas' are female. It is not difficult to imagine that such a structure might have come from an underwater coral reef. Liverworts, like lichens, are susceptible to pollution, notably car fumes and the burning of fossil fuels.

Overleaf Pellia *Pellia epiphylla*

This liverwort is very common, not just here, but can probably be found in any damp places throughout the country. Here it grows in profusion along the banks of the stream in Pope's Wood and anywhere else where the ground remains damp.

Lichens

Survey also carried out around house, walls and orchard

Scientific name	Comments
Amandinea punctata	Grows on living bark of trees
Arthonia radiata	Grows on living bark of trees
Aspicilia calcarea	Basic/limestone rock
Buellia aethalia	Grows on acidic rocks containing silica
Buellia ocellata	Grows on acidic rocks containing silica
Caloplaca citrina	Grows on acidic rocks containing silica
Caloplaca flavescens	Grows on acidic rocks containing silica
Candelariella aurella	Grows on acidic rocks containing silica
Candelariella vitellina	Found in orchard
Chrysothrix candelaris	Oak
Cladonia chlorophaea	Orchard trees
Cladonia coniocraea	Found on any dead wood
Cladonia fimbriata	Slender pixie-cups – dead wood, posts
Cladonia polydactyla	Dead wood, posts etc
Cladonia ramulosa	Grows on acidic rocks containing silica
Cliostomum griffithii	Occurs on Pine, Alder, Birch, Oak
Enterographa crassa	Grows in shade on hardwood trees
Evernia prunastri	Mainly on Oak – see illustration p124
Flavoparmelia caperata	See illustration p124
Graphis elegans	On bark of various hardwood trees
Graphis scripta	See illustration p125
Hypogymnia physodes	Mainly on Oak – see illustration p124
Hypogymnia tubulosa	Orchard tree
Hypotrachyna revoluta	Grows on living bark of trees
Lecanactis abietina	Mainly found on Oak
Lecanora albescens	Basic/limestone rock
Lecanora campestris	Grows on living bark of trees
Lecanora chlarotera	Various trees – see illustration below
Lecanora compallens	Found in orchard
Lecanora conizaeoides	Common on bark including conifers
Lecanora crenulata	Grows on acidic rocks containing silica
Lecanora dispersa	Grows on acidic rocks containing silica
Lecanora expallens	Oak
Lecanora muralis	Grows on acidic rocks containing silica
Lecanora orosthea	Grows on acidic rocks containing silica
Lecidella elaeochroma	Grows on living bark of trees
Lecidella stigmatea	Grows on acidic rocks containing silica
Lepraria incana	Grows on trees and rocks
Lepraria lobificans	Grows on acidic rocks containing silica
Leproloma vouauxii	Grows on acidic rocks containing silica
Melanelia fuliginosa ssp glabratula	Ash
Melanelia subaurifera	Oak
Parmelia saxatilis	Grows on trees and siliceous rock – called 'Crottle'
Parmelia sulcata	Frequent on Oak – see illustration p124
Parmotrema chinense	Frequent on Oak
Pertusaria amara	Various trees and walls
Pertusaria hemisphaerica	Frequent on Oak
Pertusaria hymenea	Oak
Pertusaria pertusa	See illustration p125
Phlyctis argena	Found in orchard
Physcia adscendens	Grows on living bark of trees
Physcia caesia	Grey Warted Lichen on acidic rock and trees
Physcia semipinnata	Found on trees in orchard
Physcia tenella	Found on trees in orchard – see illustration p124
Protoblastenia rupestris	Basic/limestone rock
Psilolechia lucida	Grows on acidic rocks containing silica
Punctelia borreri	Orchard trees
Punctelia ulophylla	Found on pear tree
Pyrrhospora quernea	Oak
Ramalina farinacea	Ragged Mealy Lichen – grows on various bark
Ramalina fastigiata	Orchard trees
Scoliciosporum chlorococcum	Dead wood, posts etc
Tephromela atra	Grows on acidic rocks containing silica
Usnea cornuta	Orchard trees
Usnea subfloridana	Bearded Lichen – found on orchard tree
Verrucaria nigrescens	Basic/limestone rock
Xanthoria parietina	Common Yellow Lichen – grows on bark
Xanthoria polycarpa	Orchard trees

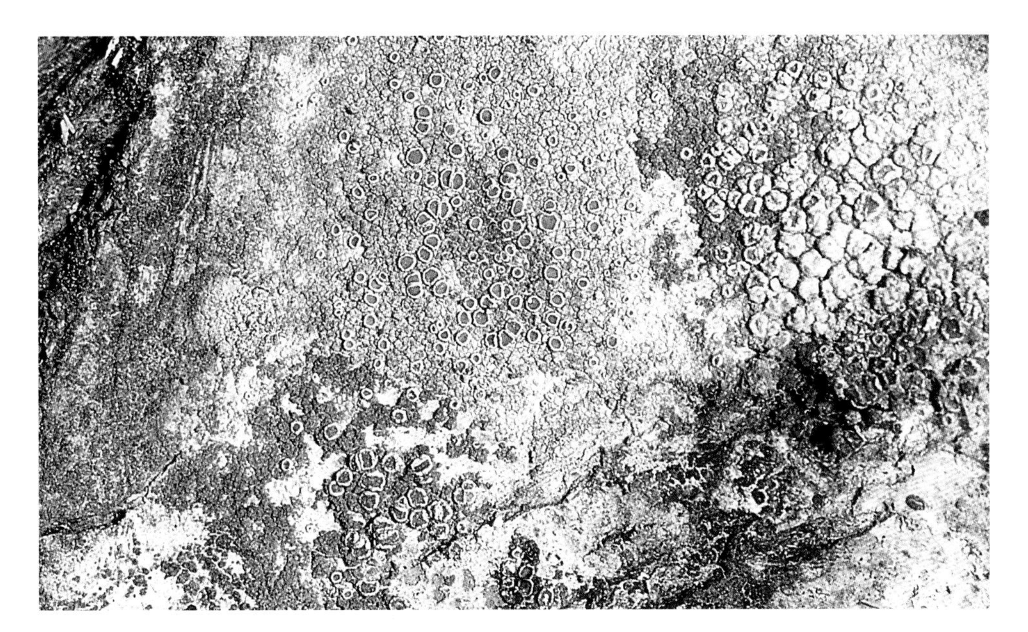

Lecanora chlarotera

This lichen grows on many trees, but this one was photographed growing on an upper branch of one of the hybrid poplars – the 'infamous' poplars growing in the western end of Pope's Wood and elsewhere, most of which are now doomed.

Physcia tenella

This little lichen was photographed on an apple twig. It shows the fruiting bodies that are only 1–2 mm across, so this picture is hugely enlarged. Lichens are rarely confined to a single species of tree but orchards are often conducive to many lichens since they are in the open and are therefore light and sunny. I wonder how many lichens could be found up in the canopy of the woods?

A mixture

This picture shows Evernia prunastri, *displaying the white underside of its leaves. It also shows the Turkeytail fungus (*Trametes versicolor*), as well as the lichen* Flavoparmelia caperata *at the very bottom. The latter grows on oaks.*

Evernia prunastri, Parmelia sulcata and *Hypogymnia physodes*

Three lichens are growing on this little Oak branch. The most prominent is Evernia prunastri, *but intermingled and below (not easily identifiable in the picture) are* Parmelia sulcata *and* Hypogymnia physodes. *The last is called Puffed Shield Lichen. There is as yet no definitive list of common names, so it may have other names as well.* Parmelia sulcata *is sometimes called Nettle Shield Lichen.* Evernia prunastri *can be distinguished from the many other stag-like lichens by the white on the undersides of its leaves.*

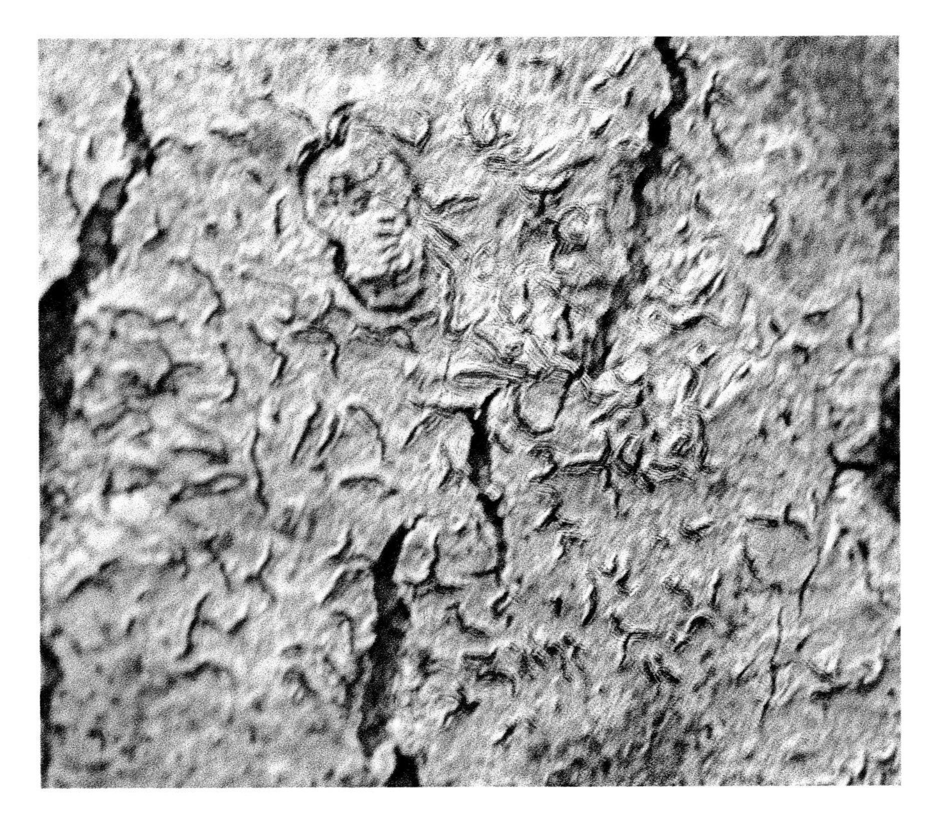

Graphis scripta

Graphis scripta *is widespread here, growing on smooth barks, mainly young Ash and Hazel. Some grow in very large patches reaching several metres up tree trunks. Viewed from a distance they look merely like a grey-green patch on the trunk; a closer look reveals some of the detail. However, much enlarged, they appear to me like static, minute black worms! These are in fact the fruiting bodies. Many lichens are very similar and can only be identified by various chemical tests. One of the 'look-alikes' of this lichen is* Graphis elegans.

Pertusaria pertusa

If the fruiting bodies of Graphis scripta *look like worms, this lichen has fruiting bodies that are like 'salt and pepper pots' – a term actually used by Jacqui! This lichen is widespread on Oak, Hornbeam and Alder.*

Mammals

Scientific name	Common name	Comments
Insectivora	Insectivores	
Sorex araneus	Common Shrew	Native; widespread
Sorex minutus	Pygmy Shrew	Native; occasionally in great numbers
Talpa europaeus	Mole	Native; widespread
Rodentia	Rodents	
Apodemus flavicollis	Yellow-necked Mouse	Native; widespread
Apodemus sylvaticus	Wood Mouse	Native; see illustration p131
Clethrionomys glareolus	Bank Vole	Native; widespread
Micromys minutus	Harvest Mouse	Native; see illustration p130
Microtus agrestis	Field Vole	Native; widespread
Mus musculus	House Mouse	Native; widespread
Muscardinus avellanarius	Hazel Dormouse	Native; several nests in Owl's and Pope's Wood
Rattus norvegicus	Brown Rat	Introduced Middle Ages; widespread nuisance
Sciurus carolinensis	Grey Squirrel	Introduced 19th century; widespread pest
Lagomorpha	Rabbits and Hares	
Oryctolagus cuniculus	Rabbit	Purposefully introduced prior to the year 1016; widespread pest
Carnivora	Carnivores	
Meles meles	Badger	Native; see illustration p127
Mustela erminea	Stoat	Native; numbers unknown
Mustela nivalis	Weasel	Native; numbers unknown
Mustela vison	American Mink	Released into the wild 1950s; destructive pest
Vulpes vulpes	Red Fox	Native; see illustration p129
Artiodactyla	Even-toed Ungulates	
Capreolus capreolus	Roe Deer	Native; widespread
Cervus nippon	Sika Deer	Introduced; less frequent than other two deer
Dama dama	Fallow Deer	Introduced by Normans; see illustration p130
Chiroptera	Bats	
Eptesicus serotinus	Serotine	See illustration p133
Myotis daubentoni	Daubenton's Bat	Can be seen swooping over Shaw Wood pond
Nyctalus noctula	Noctule	Could roost in woodpecker holes; one of the largest bats
Pipistrellus pipistrellus	Common Pipistrelle	The smallest bats found in the UK, see illustration p132
Pipistrellus pygmaeus	Soprano Pipistrelle	Only recently classified as a different species
Plecotus auritus	Brown Long-eared Bat	See illustration pp132 and 133

Currently there are only twenty-seven native mammal species in the UK – marine animals and bats excepted. This means that our small woodland supports more than fifty per cent of our native terrestrial mammals.

Unfortunately, other mammals have been introduced over the years, many of which (like plants) have become great pests and now upset the natural balance. The reintroduction of the wolf and the beaver into the British Isles could do much to redress this imbalance, but I fear any moves in that direction (some are afoot) will meet with tough resistance both from powerful vested interests and through ignorance of the true nature and worth of these animals. The wolf must surely be the most persecuted and misunderstood animal in existence.

As far as our woods and wildlife are concerned, the prime pests are the rabbit, the grey squirrel and the mink. Rabbits were introduced in the twelfth century and at least had the saving grace that they were brought in for food for the local people, whereas the first grey squirrel was introduced from North America as late as 1876, merely as a fun novelty.

Rabbits and grey squirrels may be cute to observe, but both are real pests to trees and plants. Both will gnaw and strip the bark off trees, especially young ones, and even if the tree does not die immediately, the open wounds leave them prone to attacks from beetles, other insects and diseases in subsequent years. Grey squirrels can kill quite large trees by wantonly stripping the bark, and their favourite targets here are Beech and Hornbeam. Unfortunately, it does not stop there: rabbits will raze many ground plants and often those that are least widespread. At least they graze for food; grey squirrels on the other hand will damage anything they fancy for no apparent reason. They tear up birds' nests, with eggs or young present, and nip buds, flowers and fruit willy nilly, leaving them strewn all over the ground. I am afraid we have no compunction about shooting or humanely trapping either of these mammals. As we are surrounded by farmland, which squirrels rarely cross, we have had a very small measure of success in keeping the squirrel population under control, despite the fact that we are its only predator.

Myxomatosis and foxes both contribute to the rabbits' death toll (indeed I can imagine that the foxes are getting quite bored with a constant easy-prey diet of rabbit meat), but despite this, the population is still way above what nature would consider a good balance.

Photograph © Alan Shears

Young badger cubs *Meles meles*

Much has been written about the devastating effect of deer on woodland plantations, but I am convinced this is only true where plugs or saplings are actually planted rather than self-generated. Had we not protected the trees that were planted in these woods after The Storm, I am sure very few would have survived. However, having observed trees self-generating over the last sixteen years, I know that deer or rabbits have only marginally attacked them. Scientists have not yet come up with a good answer as to why this should be, but I am convinced Nature must have devised a natural protection for plants that germinate and grow where and when they choose. I have just discovered that a book apparently dealing with this phenomenon is to be published about the same time as this one.

We have the three species of deer but it is the roe and fallow deer which often appear in great numbers. They obviously cause some damage, and in the last few years this has increased. Here they seem particularly fond of Guelder Rose and Spindle Tree, but these are both plants that regrow by coppicing or suckering so they are rarely completely destroyed. I have observed two Spindles side by side: one has been left alone since it was planted, but the other is annually grazed to within a few inches of the ground. The Alder Buckthorns that were planted in 1990 were left alone until 2002. Many have now been badly damaged.

Bucks and stags trying to clean their antlers prior to rutting can cause considerable damage to a tree, but I am glad to say they nearly always go for small conifers, of which we have too many anyway! It is difficult to tell whether deer graze some of the woodland plants, but I have never noticed any damage to either Bluebell or Wood Anemone of which there must be millions growing densely. It would be obvious if they had been disturbed. A neighbour tells me he has seen a muntjac deer in his garden. I trust he is mistaken as these small deer (also imported) can be as bad or worse than rabbits. So far, I have not observed any.

The most reviled of all mammals must be the rats. Brown and Black Rats both arrived here from ships in the Middle Ages. I am uncertain whether they upset the balance of nature here in these woods (they certainly do so in other parts of the world), but they do carry diseases that can be passed on to humans. Yet our clinically antiseptic but wasteful, modern society seems to contrive to do everything that allows them to proliferate even more. It is said that no one is ever more than a metre or so way from a rat! Brown rats persist in our barn, especially in wet winters. We constantly try to poison them, but they are wily creatures and ignore the poison in their runs unless mixed with expensive minced beef! I also suspect that they have become immune to some of the poisons. They obviously come for the chicken food but, oddly, I have never known them to attack small chicks or take eggs (they seem to leave these for the crows which fly even into the barn!).

The mink is able to prey on almost anything, in or out of water and even up trees. Fish, birds, waterfowl and many mammals are all part of its daily diet. Misguided animal activists released the mink from fur farms (which most people disapproved of anyway) during the second half of the twentieth century. I wonder if today's farmers and gamekeepers would not rather opt for the now rare native polecat (*Mustela putorius*) and pine marten (*Martes martes*) rather than the mink?

Having maligned some of our mammals, I hasten to add that I love most of them. The grace and elegance of the deer is wonderful to watch, and even if the damage they do is occasionally annoying, I would not be without them. I had only been here two years when the whole of the vegetable garden was made deer and rabbit proof. You have no idea how partial deer are to strawberries, artichokes and much else!

We have many badgers. There are at least two setts, but badgers move around constantly, especially in summer when they seem to interchange with fox lairs and rabbit warrens, so that it is difficult to tell if there are any more. Badgers can be viewed most evenings at the top of Badgers' Wood, and visitors have often seen seven or eight together. The woods are littered with their trails.

The apparent total absence of hedgehogs (*Erinaceus europaeus*) is a mystery. I certainly have never seen one in twenty-five years. Maybe the presence of so many badgers could account for this?

I have many recollections about foxes, but as they are territorial (each dog claiming one to ten square kilometres), I am sure they are mostly members of the same family. The local fox hunt has always been barred from my woods (not without difficulty, and occasionally we are still overrun by hounds). It often seemed as if all the foxes in the neighbourhood ended up here when a hunt was in progress. Quite apart from any moral or ecological objections I may have to the hunt, I find the claim that they protect farming livestock tenuous, to say the least. It is not difficult to protect chickens and other birds from foxes, which only leaves lambs that are in the open vulnerable to predation.

I used to keep a handful of ewes, which lambed most years. I kept a few bales of hay on wooden pallets straddling a dry ditch just outside the orchard where the sheep grazed. One late spring, I found my dog had cornered a fox cub not much older than the one pictured opposite. Despite all the barking and snarling, both dog and fox were untouched. I called off my dog, and the cub ran straight into the ditch under the plastic-covered bales of hay. I discovered that the foxes had turned this into a ready-made earth. As far as I could make out there were three youngsters. How both my dog and I had not discovered them before is a mystery (maybe they had only recently moved in), but the fact remains that no lambs, either that year or any other, were ever harmed.

I have found shrews and various mice in my compost heaps. To some people's horror, I have a soft spot for mice, as well as shrews and voles. The latter can damage shrubs and trees, but as there are plenty of predators here (foxes, owls and other birds of prey) a healthy balance is maintained. Mice, even house mice, can be intriguing, and if there were space, I could relate many a story about them. Instead, I hope the pictures on pages 130-131 beguile you.

Bats are abundant here, but it was not until the survey was made over several nights that I became aware that we have more than one species. Their presence is largely due to the various ponds that spawn many insects, including the irritating ones. Bats, like wolves and snakes, still have a bad public image fostered over centuries, but how people in this day and age can moon over squirrels and yet still have a horror of these silky, endearing little creatures is hard to imagine. Maybe some people are unaware that just a few bats can consume more than a million exasperating mosquitoes and midges in one night – that, if nothing else, should endear them to us!

We found evidence of bats (brown long-eared) in only one building, so where they roost is something of a mystery since so many of the really old trees were destroyed in The Storm. However, we now have quite a number of bat boxes on trees near ponds. Roger Jones (who also kindly installed the bat boxes) mainly carried out the survey, and I also have to thank Caroline Stone for allowing me to photograph some of her bats. It seemed extraordinary that these night creatures were totally unphased by photographic flashes.

Photograph © Alan Shears

Red fox *Vulpes vulpes*

Note how the young cub still has a round face with a short snout. Adults' faces become long and pointed.

Photograph © Alan Shears

Harvest mouse *Micromys minutus*

Many people have an inborn fear of rats and mice, even to the point of paranoia. I can understand the former, but it is difficult to imagine anyone being in terror of the tiny and enchanting harvest mouse. I am glad to say that they are quite numerous in the woods, and also in the garden, which is not tended or weeded very often so they have an ideal habitat. I frequently find their abandoned nests (tightly woven bundles of grass and leaves) and occasionally one that is still inhabited – but never when I have a camera! Today, they are not as numerous as they once were, having largely lost their habitat of traditional cereal farm fields. They now have to make do with hedges and woods. Harvest mice are really tiny and weigh only around five grams.

Fallow deer *Dama dama*

I am very disappointed that I never managed to photograph the many deer that roam these woods. I am afraid it was always a question of time and, in my case, patience. Their hearing and sense of smell are phenomenal. Even when five magnificent stags appeared on the lawn one evening, I knew it would be hopeless – the merest sound of an opening window would have them bounding off. The young buck in the picture also appeared outside a window, the one where I had my computer. I managed to fetch the camera and took this not very good picture through the glass. He was aware of my presence and stood quite still (posing?) until I tried to open the window – then he was gone in a flash. He is standing at the edge of Shaw Wood; note the rhododendrons in the background, which have since been removed.

Photograph © Alan Shears

Wood Mouse *Apodemus sylvaticus*

I think I can say fairly that the wood mouse is more common here (including in the house) than the ubiquitous house mouse. It is a little larger than the harvest mouse and much more gregarious. I had one pair who persistently followed the sack of bird food. No matter where I moved it – from the garage to the hall, larder or kitchen – they found it and gnawed through it, scattering seed everywhere. In the end I stored the bird seed in plastic tubs with tight-fitting lids, but I still have a lid pocked with teeth marks where they tried to prise it open! Foiled at last, they had to content themselves with climbing up on to the bird table or, better still for my entertainment, leaping on to it from a nearby shrub. In case you think I am making this up or that there must have been numerous pairs, I could always recognise these two – the male had a large chunk missing out of one ear. Mice and voles are an important part of the diet of owls and other birds of prey, but I hope these two wood mice managed to live out their lives naturally.

Brown long-eared bat
Plecotus auritus (see also page 133)

These bats are quite distinctive with large, curled, ram's-horn-like ears almost as long as their bodies. They are medium sized, but only a little larger than the pipistrelle. Like all the other British bats, their main food source is insects and beetles, but unlike the others they often fly close to the ground or even land to catch their prey. This was the only species we found in any building.

Pipistrelle *Pipistrellus pipistrellus*

This picture, like the others on these pages, is much enlarged. Pipistrelles are Britian's smallest bats. Watching them in flight can be deceptive as their wingspan is wide (18–24cm), but their bodies are tiny, 3–5cm, and they weigh less than ten grams. We have quite large numbers of pipistrelles that fly around the ponds at night in search of insects. They are said to prefer buildings for roosting, but Roger could not find any evidence of colonies in either the house or outbuildings. I did, however, find a dead one some years ago in the barn, so maybe they do roost there. The barn has many holes (small birds nest there) that are impossible to check. Only a matter of a month or two after Roger had installed the bat boxes, he found one in Shaw Wood already inhabited by Pipistrelles. They are the first to emerge, when it is still fairly light, and it is wonderful to watch them weaving and pirouetting in the twilight.

Serotine *Eptesicus serotinus*

The Serotine is about twice the size of the Pipistrelle. Like all British bats, it feeds on insects, often swooping down to the ground to seize them. Again, no evidence of their roosting sites could be found in any buildings – perhaps they have found some old tree holes (deserted woodpecker nests of which we have many?) which, after all, must have been their natural homes when forests formed large, endless tracts. This particular bat is called Sammy and lives in Caroline's hospital. Alas, he will remain there for the rest of his life because, although he can fly, his flight is not strong enough to survive in the wild. His rather fierce-looking baring of teeth is nothing to do with fear or annoyance. He is, in fact, echo-locating – a process similar to radar. Each bat species emits a different signal pattern. The signals sent by an echo-locating bat can be picked up by electronic bat detectors, enabling humans to identify the various species. When a bat flies off in the dark, it has to be able to locate objects and obstacles around it. Echo-location allows a bat to avoid obstacles, and find moths and other insects to eat. But there is quite a high-tech battle going on at night: some moths can hear a bat a long way off so they can hide or take evasive action. Some have learnt to emit pulses of their own, advertising their unpalatableness. But bats have learnt to take counter-measures! They also have to be able to avoid becoming the prey of bigger predators like owls, which would relish a bat. And so the carefully balanced struggle between predator and victim goes on.

Brown Long-eared bat

Note

Bats are a highly protected species in Europe. It is an offence to harm or disturb them in any way. It is also an offence to photograph them without a permit. The pictures on these two pages are all of rescued bats in Caroline's hospital.

Birds

When I first moved here, it was almost impossible, even for me, to sleep through the dawns of spring. The chorus was like a full-blown symphony orchestra, some of it emanating from right outside my window. This 'dawn music' is still here, but I am sure the decibels have diminished over the years. Is this down to the gradual decline of many birds in the last thirty years, or have I become accustomed, or am I just not hearing so well?

Also, when I first came here nearly a quarter of a century ago, I did not believe in feeding wild birds – or, indeed, any wild animals. Nature should take its course – survival of the fittest and all that. This is all very well in large nature reserves, or the few still wholly wild places in the world where there is still a good natural balance, but here

Common name	Scientific name	Comments
Grebe family	*Podicipedidae*	
Little Grebe	*Tachybaptus ruficollis*	Infrequent visitor to Shaw Wood pond
Heron family	*Ardeidae*	
Grey Heron	*Ardea cinerea*	Constantly fish in Shaw Wood pond
Duck family	*Anatidae*	
Canada Goose	*Branta canadensis*	Occasional visitors; have bred on large pond
Mallard	*Anas platyrhynchos*	Resident; breed on Shaw Wood pond and elsewhere. See illustration p140.
Mandarin Duck	*Aix galericulata*	Occasional visitors
Tufted Duck	*Aythya fuligula*	Have visited Shaw Wood pond
Hawk family	*Accipitridae*	
Sparrowhawk	*Accipiter nisus*	Occasional
Falcon family	*Falconidae*	
Hobby	*Falco subbuteo*	Seen during autumn migration
Peregrine	*Falco peregrinus*	Seen prospecting overhead
Kestrel	*Falco tinnunculus*	Regular; probably breeds. See p138.
Pheasant family	*Phasianidae*	
Pheasant	*Phasianus colchicus*	Many present
Rail family	*Rallidae*	
Moorhen	*Gallinula chloropus*	Breed on ponds in Shaw Wood and Morgan's Strip
Plover family	*Charadriidae*	
Lapwing/Peewit	*Vanellus vanellus*	Seen overhead
Waders		
Woodcock	*Scolopax rusticola*	Present several years after 1987 but now a winter visitor
Gull family	*Laridae*	
Black-headed Gull	*Larus ridibundus*	Seen overhead
Lesser Black-backed Gull	*Larus fuscus*	Seen overhead
Herring Gull	*Larus argentatus*	Seen overhead
Pigeon family	*Columbidae*	
Collared Dove	*Streptopelia decaoto*	One pair comes to bird table and vegetable garden
Stock Dove	*Columba oenas*	Occasional
Woodpigeon/Ring Dove	*Columba palumbus*	Always present and breed
Cuckoo family	*Cuculidae*	
Cuckoo	*Cuculus canorus*	Regular summer visitor, even in the house!
Owl family	*Tytonidae and Strigidae*	
Barn Owl	*Tyto alba*	Occasional visitor from surrounding farmland, may breed
Tawny Owl	*Strix aluco*	A few pairs breed. See p141.
Little Owl	*Athene noctua*	Occasional visitor; may breed
Long-eared Owl	*Asio otus*	Bred one year in Badgers' Wood but not seen since
Nightjar family	*Caprimulgidae*	
Nightjar	*Caprimulgus europaeus*	Regularly heard, but not every year
Swift family	*Apodidae*	
Swift	*Apus apus*	Often seen in numbers before migration
Kingfisher family	*Alcedinidae*	
Kingfisher	*Alcedo atthis*	Often visits Shaw Wood pond; may have bred. See p139.
Woodpecker family	*Picidae*	
Great Spotted Woodpecker	*Dendrocopos major*	Common breeding resident. See p143.
Lesser Spotted Woodpecker	*Dendrocopos minor*	Possible recent sighting in Butler's Wood
Green Woodpecker	*Picus viridis*	Common breeding resident
Lark family	*Alaudidae*	
Skylark	*Alauda arvensis*	Nests in surrounding farmland

Cuckoo *Cuculus canorus*

I do not think I had ever seen a cuckoo until this one turned up in my living room and, like the woodpecker (see p143) went straight for the cacti. By the time I found him, he was covered in spikes which I eventually managed to remove. I even managed to take this picture, the bird in one hand and the camera in the other! Inevitably he was stressed, and as it was late evening in autumn (Chris reckons he may have been on his way south), he spent the night in a box with water and food. He hopped out next morning and flew off, apparently none the worse for his prickly experience.

Common name	Scientific name	Comments
Swallow family	*Hirundinidae*	
Swallow	*Hirundo rustica*	Regular summer visitor. See p140.
House Martin	*Delichon urbica*	Regular summer visitor
Wagtail family	*Motacillidae*	
Pied Wagtail	*Motacilla alba*	Often present in Barn area
Grey Wagtail	*Motacilla cinerea*	Occasionally seen in Barn area
Starling family	*Sturnidae*	
Starling	*Sturnus vulgaris*	Breeding resident. See p137.
Accentor family	*Prunellidae*	
Hedge Sparrow/Dunnock	*Prunella modularis*	Fairly widespread; breeds
Wren family	*Troglodytidae*	
Wren	*Troglodytes troglodytes*	Always present and breeds. See p140.
Thrush family	*Turdidae*	
Blackbird	*Turdus merula*	Always present and many breed
Song Thrush	*Turdus philomelos*	Always present and many breed. See p142.
Redwing	*Turdus iliacus*	Winter visitor
Fieldfare	*Turdus pilaris*	Winter visitor
Mistle Thrush	*Turdus viscivorus*	Present in numbers and breeds
Nightingale	*Luscinia megarhynchos*	Song heard several times in twenty years
Robin	*Erithacus rubecula*	Present in numbers and breeds
Warbler family	*Sylviidae*	
Chiffchaff	*Phylloscopus collybita*	Common breeding summer visitor
Goldcrest	*Regulus regulus*	Breeding resident
Blackcap	*Sylvia atricapilla*	Common breeding summer visitor. See p137.
Firecrest	*Regulus ignicapillus*	Possible occasional visitor
Willow Warbler	*Phylloscopus trochilus*	Breeding summer visitor
Flycatcher family	*Muscicapidae*	
Spotted Flycatcher	*Muscicapa striata*	Autumn migrant
Tit family	*Paridae*	
Coal Tit	*Parus ater*	Present in large numbers and breeds
Blue Tit	*Parus caeruleus*	Present in large numbers and breeds
Great Tit	*Parus major*	See illustrations pp139 and 141
Long-tailed Tit	*Aegithalos caudatus*	Present in numbers and breeds
Marsh Tit	*Parus palustris*	Present in numbers and breeds
Nuthatch family	*Sittidae*	
Nuthatch	*Sitta europaea*	Present in large numbers and breeds
Creeper family	*Certhiidae*	
Treecreeper	*Certhia familiaris*	Fair numbers present and breeds
Bunting family	*Emberizidae*	
Yellowhammer	*Emberiza citrinella*	A few hold territories in woodland edges and surrounding hedges
Finch family	*Fringillidae*	
Brambling	*Fringilla montifringilla*	Occasional visitor
Chaffinch	*Fringilla coelebs*	See p136
Bullfinch	*Pyrrhula pyrrhula*	Small numbers present and breed
Goldfinch	*Carduelis carduelis*	Some years large numbers appear
Greenfinch	*Carduelis chloris*	Present in numbers and breeds
Siskin	*Carduelis spinua*	Winter visitor in small numbers
Sparrow family	*Passeridae*	
House Sparrow	*Passer domesticus*	Several pairs breed in Barn area
Crow family	*Corvidae*	
Carrion Crow	*Corvus corone corone*	Present in too great numbers
Jackdaw	*Corvus monedula*	Sometimes breed in chimney pots
Jay	*Garrulus glandarius*	Present, but not in large numbers
Magpie	*Pica pica*	Too many present
Rook	*Corvus frugilegus*	Occasional visitors

in the UK nothing is any longer totally wild or natural. The more I thought about it, the more I came to the conclusion that I might, if not increase the diversity, at least maintain and increase the numbers of some species. In fact, the situation seemed similar to plants. If there is insufficient space for them elsewhere, it seems sensible to endeavour to cram in as much as possible into what space is available. We had enough nesting habitats, but it was uncertain if the food supply was sufficient despite the fact that many trees and shrubs were planted with bird food in mind.

So, I acquired two feeders, and we also made a table. I would not now be without them – the birds entertain for hours. Today we estimate that there could be several hundred who come to feed. All the small birds, like finches and tits, come in great numbers as well as many larger ones. The nuthatches, I soon discovered, are the bullies, scattering everyone in their wake. They also have a pecking order among their own kind. The greenfinches are a riot and even bring their families – the babies crouching with trembling wings, begging for food, whilst the parents stand aloof as if to say, 'can't you see there is food all around you'! Larger birds, such as blackbirds, tolerate the smaller ones darting around them. Collared Dove are so large they take up most of the table anyway. I could write reams about the beguiling antics of the birds.

However, not all birds are either enchanting or beguiling: for instance, members of the crow family, in particular, Carrion Crow. They live here in large numbers – wily and intelligent and all part of the natural scene. However, they have few predators and appear to have a monopoly on marauding amongst the other birds – perhaps the increasing presence of raptors will help. I could hardly contain my fury when one of them destroyed a Song Thrush's nest near the house. I doubt whether she nested again.

I welcome the increasing numbers of raptors, but do they not create another dilemma? We hear constantly about the appalling decline of many bird species. The main reasons given are loss of habitat, the inordinate number of pet cats, pollution and the use of chemicals, all of which inevitably upset any natural balance. Small birds loom large in the diet of many birds of prey, so are we creating yet another hazard for many declining small birds when we assist birds of prey?

I am also ambivalent about Grey Heron. They are great, primeval-looking birds and I am really fond of them. However, despite the presence of hundreds of Roach in the large pond, herons have virtually foiled all my attempts to introduce any other species of fish. Two large black Carp survive (I hope they are now too large for the ever patient heron) as well as a handful of Golden Orf. I must have put dozens of Golden Orf into the pond over the years, but they have never bred. Maybe I should try introducing the native Tench?

There was an unusual incident one year involving crows and the heron. One day pandemonium broke out on the pond. I frequently hear a cacophony there and it normally heralds the arrival of a gaggle of Canada Geese, but this was different. I rushed out, but everything went quiet. The next day I heard the same thing, but this time I was more cautious in my approach. It turned out that quite a number of crows and one heron were harassing each other with unholy fury. I can only surmise that the latter thought he was protecting his fish, and the crows were fearful about their next meal of ducklings. It was astonishing to see one heron holding his own against a murder (very apt!) of crows. I never discovered who won the skirmish, but nothing similar has occurred since, and both species still come to the pond.

I thought I had acquired a little knowledge about birds over the years, but when it came to researching them for this book, my total ignorance became woefully apparent. I still cannot identify many birds, and I certainly could not match many calls and songs to particular species. So, as with the surveys in the other chapters, I turned to an expert – Chris Gent, who spends all his spare time roaming the world with his wife searching for more and more birds. They spent many days and weekends here over two or three years surveying the woods. Chris discovered many species I did not even realise existed here. In the case of the Nightingale and the Woodcock, he had to rely on my word. The former's song is unmistakable, and the latter were definitely here several years after The Storm. At one point in 2003, Roger Jones, the bat expert, glimpsed what appeared to be a Lesser Spotted Woodpecker at the top of Butler's Wood. Chris never confirmed the sighting of this very shy bird but it is included in the survey. He says this is an ideal habitat and some should be present.

For some of the photographs I turned to Alan Shears. I realized it would take many months to obtain the wonderful photographs he already had. This chapter would not have been possible without either Chris or Alan.

Photograph © Alan Shears

Chaffinch *Fringilla coelebs*

"Are you still all hungry?" Chaffinches are now the second most common bird in Britain and also one of the most colourful. The species name coelebs, *meaning bachelor, originates from chaffinches' habit of forming large flocks of one sex in winter.*

Starling *Sturnus vulgaris*

I can remember the time when there were swarms of starlings, chattering and wheeling around. Sadly this no longer seems to occur. However, several pairs at least still nest here, and I have found them under the shiplap in the barn and sheds. I seldom see them at the bird table or feeders – maybe none of the food is to their liking.

Photograph © Alan Shears

Blackcap *Sylvia atricapilla*

"Don't you think they have eaten enough today?" These wonderful birds have a stupendous song quite out of all proportion to their tiny size. I am glad they are so numerous here.

Photograph © Alan Shears

Kestrel *Falco tinnunculus*

We have put up a Kestrel nesting box in the woods, but so far it appears to be inhabited solely by Pigeons. Kestrels certainly hunt here, but I was very surprised to see one on the bird table outside the kitchen window! Although a small bird of prey, he looked enormous compared to the usual visitors. He just gazed around with that sharp turn of the head, and I immediately feared for the small birds (one becomes quite possessive and protective!), but the Kestrel just flew off only to return the next day. He was actually after mice (who often come scavenging bird-seed) because he suddenly swooped to the ground and was gone in a flash with something! As far as I know he has not been back – yet.

Photograph © Alan Shears

Kingfisher *Alcedo atthis*

Kingfishers are often seen flashing round the pond in Shaw Wood, which is full of the right food. We have literally hundreds of Roach. I cannot say with certainty that kingfishers nest here, but I am fairly sure they do. I found a young bird one day that had flown to its death by crashing into a large window.

Great Tit *Parus major*

I get quite used to birds in the kitchen. They frequently come in and sit on the window sill, especially when the feeders are empty. I am sure they are demanding that I get up and produce more food. Of course, they often forget the way out and frequently end up in the most improbable places – swinging on the lights, balancing on cups or demolishing my plants! This Great Tit, however, turned up in my bedroom where I actually had a camera handy, and this is the result.

Mallard *Anas platyrhynchos*

These ducks often appear to be extraordinarily stupid parents. They allow their young to become easy prey for the crows, remaining on the large pond without any warnings of danger, unlike the moorhens whose high-pitched 'rattle' sends the young scuttling for shelter. Not, however, this mother Mallard. She appeared one day with fourteen chicks on the tiny pond in the Barn area, only to disappear two days later, and then she turned up on the pond in Shaw Wood. This continuous switching of venues went on for quite some time until the chicks were much more able to look after themselves. Nevertheless many were lost to the crows, but I think at least five or six survived – a very good percentage.

Swallows and wrens *Hirundo rustica* and *Troglodytes troglodytes*

The picture in the middle shows baby swallows in the nest. After twenty years of nesting in our garage, the swallows failed to return in 2001. It was a devastating blow, but I heard later we were not alone – a number of swallows had apparently deserted Sussex to nest further north. Another sign of climate change? In 2001, the wrens – no longer harassed by the swallows – nested in the garage. The following year, assured that the swallows would not return, they built this mossy green second storey on top of the same nest pictured in the middle. I never managed to photograph them with their heads peeping out like the swallows – they always stayed well hidden, and then suddenly they were gone.
Close to the garage there are three Black Mulberry and although I do not believe they are the only culprits, it is the wrens that always manage to strip the bushes bare before we have a chance to pick any of this delicious fruit.

Photograph © Alan Shears

Tawny Owl *Strix aluco*

Numerous Tawny Owl inhabit these woods – at least it sounds like it when one hears their conversations at night. Also, at dusk, they can often be seen in or around Shaw Wood. Chris tells me they must have been responsible for ousting the Long-eared Owl that nested one year in Badgers' Wood.

Great Tit *Parus major*

For many years I had successfully foiled various birds' attempts to nest in this drain. Not only was it hazardous for them, but it saved me having to put up notices in rooms: "do not turn on water – birds nesting!" However, a pair of great tits sneakily flattened the wire netting and had built a nest before I realised what was happening! As you can see, it was quite a squeeze for the parents to get in and feed the young. Most of my concentrated birdwatching time takes place around the house and bird feeders. When I am in the woods I am usually concentrating on all kinds of things and do not take in too much detail about the birds. Several people tell me that many birds (maybe hundreds) must come from quite a long way to the feeders. It is certainly true that if there is a sudden movement or noise, swarms of birds take off from the bushes and trees around the bird feeders.

Photograph © Alan Shears

Song Thrush *Turdus philomelos*

We are lucky to have several pairs of Song Thrush as well as Mistle Thrush. Neither come often to the bird table where I put out special food, but blackbirds and robins are regular visitors. However, I often find song thrushes trapped inside the fruit cage in the vegetable garden – not a bad place to spend a night or two! Any snails I find are collected and scattered on the lawn where the thrushes are not slow in finding them.

Photograph © Alan Shears

Great Spotted Woodpecker ***Dendrocopus major***

Brash or just plain stupid? The Great Spotted Woodpecker have given endless entertainment, despite the fact that they ousted the Green Woodpecker from the feeders. One year they flattered me by bringing all their young family. It was an astonishing, heart-warming sight to see two adults and, I think four young circling and calling outside the kitchen window. The father was something else: three times over a short period he managed to get himself caught in the squirrel and rabbit traps! When released, he always flew away none the worse after several hours of confinement. As he obviously was not going to learn, we ceased to set the traps. Not content with that adventure, he took to flying through open doors into the living room and, like the Cuckoo, got himself entangled in the cacti! I always managed to catch him to pull out any spikes, but I have never heard a bird produce such demonic shrieks when held. The picture shows mother feeding a young son.

Grass Snakes *Natrix natrix*

Note the presence of grass snakes' eggs, as well as various sized slow worms.

Amphibians and Reptiles

Unfortunately, the number of species in this group is very small but, as there are only double the number in the whole of Britain, I cannot complain. We have vast numbers of grass snakes and slow worms, and they can be seen almost daily on my compost heaps in warm weather. It seems that these two species cohabit very successfully, producing eggs and young on the same heap, as you can see from the picture above in which there are hatched and unhatched snake's eggs.

I am delighted about the large number of slow worms, especially as they are so near the vegetable garden where I am sure they keep down the slug population. However, I am ambivalent about the grass snakes as they eat frogs and toads, but apparently only five to eight in one year. Frogs and toads seem to have decreased dramatically in recent years.

Common name	Scientific name	Comments
Amphibians		
Palmate newt	*Triturus helveticus*	*Present in all ponds except the one in Morgan's Strip*
Smooth newt	*Triturus vulgaris*	*Present in all ponds except the one in Morgan's Strip*
Common toad	*Bufo bufo*	*Fairly widespread*
Common frog	*Rana temporaria*	*Widespread*
Reptiles		
Viviparous lizard	*Lacerta vivipara*	*Widespread*
Slow worm	*Anguis fragilis*	*Present in great numbers*
Grass snake	*Natrix natrix*	*Present in great numbers*

I can remember fifteen years ago being unable to drive up our track at night due to hundreds of frogs hopping across. We have plenty of suitable habitats and use no chemicals, so the reason for their decline must lie elsewhere. For instance, in 2003, there was a large area of frog spawn in the pond in Morgan's Strip, but not one of the eggs appeared to be fertile. Common toads, never present in great numbers, have also declined. It is many years since I found any eggs.

Newts and viviparous lizards are more difficult to spot, but I am certain their populations are quite large. I admit I have only seen lizards in and around the garden, but they must be present elsewhere.

Barry Kemp, a member of the local Reptile Group, who helped with the survey, is sure we must have adders but as neither he nor I have managed to spot any, they do not appear in the final list.

Slow Worm *Anguis fragilis*

Slow worms of all shapes and sizes (and a grass snake).

Palmate newt *Triturus helveticus*

There are only six native amphibians.

Juvenile viviparous lizard *Lacerta vivipara*

There are also only six native reptiles.

Gatekeeper *Pyronia tithonus*

Butterflies and Moths

I wrote in the introduction that I would not include insects and beetles, and yet here I am with a chapter on butterflies and moths! At the nth hour, I was even persuaded to include some dragonflies! However, as they are all not only spectacular, but much enjoyed and loved by many, I decided to include them.

With a few exceptions, woodlands are not ideal for butterflies, but even here we have found almost fifty per cent of all of the species native to Britain (there are only just over fifty, plus a number of summer migrants). If you want good colonies of butterflies in your garden, you have to put up with nettles, docks, thistles and the like – weeds, as we tend to call them! Many will readily come for nectar to Buddleia and Stonecrop, but they also seek foodplants for the larvae, on which they will lay their eggs, and these plants are rarely cultivated species. Of course, there are some like the Large and Small White who may decimate your brassicas or invade your fruit crops! Hemp Agrimony, Bird's-foot-trefoil, Wild Angelica and Hogweed also attract butterflies.

We have the ideal habitat for Purple Hairstreak, but regrettably, I have only seen 'something' high in the trees! It may have been this species but I couldn't be sure. In the survey, largely carried out by myself with some help from the Butterfly Conservation Society, I have also given some of the foodplants for the larvae.

Moths are quite different: there are vast numbers endemic to Britain that, as I soon discovered, can be split into several bewildering but distinct groups – macro, micro, pyralid, plume, tortricoid and others. Together, the number of species runs into thousands, including some that are not, as yet, identified.

My moth mentor was Colin Pratt, a leading expert, especially on the moths of Sussex. We focused on the macro group, but one or two others crept in, largely because I found them so fascinating and beautiful! There is much less information available on the very small moths and, with one or two exceptions, they have not yet acquired common names.

During the latter part of 2002, I purchased a moth trap, which I put out when time permitted. The first time I opened the trap it was like gazing into Aladdin's cave – moths of all shapes and colours confronted me, from large hawk-moths to tiny fast-flying creamy-beige ones that invariably escaped and inhabited the house for days on end. The idea was that I would photograph them and Colin would try to identify them.
I say *try*, not to cast aspersions on Colin's identification abilities, but because many of the pictures were appalling, having been taken with a camera held free-hand in the most improbable locations. Nevertheless, he coped with all of them!

I quickly realised that to come even close to detecting all of the species that inhabited the woods I would have to put out the trap daily. Even if I had time for that, it would take years – a lifetime – to find them all. Nevertheless, Colin and I accumulated a fair number. Each 'trapping' produced one or two new species, even as late as November. Had I been prepared to kill the lively ones, we would have had many more, and maybe some rare ones. For instance, I found what I hoped was the Sussex Emerald but, despite keeping it in the refrigerator for two days in an attempt to subdue it for a photograph, it still escaped! It has therefore gone down as its near-twin the Common Emerald. I have given some of the foodplants of the larvae (ones that grow most commonly here) in the list, and, unless otherwise indicated, all the moths belong to the macro-lepidoptera.

Butterflies

Common name	Scientific name	Food plants/comments
Brimstone	*Gonepteryx rhamni*	Alder Buckthorn
Brown Argus	*Aricia agestis*	Cranesbill, Rock Rose
Clouded Yellow	*Colias croceus*	Clover, Vetch. Migrant.
Comma	*Polygonia c-album*	Nettle, Elm
Common Blue	*Polyommatus icarus*	Birdsfoot-trefoil, Clover.
Essex Skipper	*Thymelicus lineola*	Grasses
Gatekeeper	*Pyronia tithonus*	Grasses. See p146.
Green-veined White	*Pieris napi*	Plants in the Cruciferae family. See p148
Holly Blue	*Celastrina argiolus*	Holly, Ivy, Spindle, Gorse
Large Skipper	*Ochlodes venata*	Grasses
Large White	*Pieris brassicae*	Brassicas and Nasturtiums
Meadow Brown	*Maniola jurtina*	Grasses
Painted Lady	*Cynthia cardui*	Thistles, Burdock – Migrant. See p148.
Peacock	*Inachis io*	Nettle. See p149.
Red Admiral	*Vanessa atalanta*	Nettle. Migrant. See p149.
Ringlet	*Aphantopus hyperantus*	Grasses
Small Copper	*Lycaena phlaeas*	Dock, Sorrel
Small Heath	*Coenonympha pamphilus*	Grasses
Small Skipper	*Thymelicus sylvestris*	Grasses
Small Tortoiseshell	Aglais urticae	Nettle
Small White	*Pieris rapae*	Scourge of the vegetable gardener
Speckled Wood	*Pararge aegeria*	Grasses
White Admiral	*Ladoga camilla*	Only Honeysuckle

Speckled Wood *Pararge aegeria*

Green-veined White *Pieris napi*

Painted Lady *Cynthia cardui*

Butterflies, especially the ones that come readily to Buddleia and Stonecrop, sit quietly, wings open, often relying on their bright colouring or imitation 'eyes' to ward off predators. A good example is the Peacock's 'eyes' (right), and the bright splashes of colour on the Painted Lady (bottom p148). There are many nettles in the Barn area, and in some years we have had a glut of Red Admiral, many of them becoming trapped in the barn. Both the Red Admiral and the Painted Lady are migrants. Bright colours are also used to attract the sexes to each other.

On the other hand, camouflage, especially on woodland species, is important in evading predators. Both the Speckled Wood and Gatekeepers (page 146–147) can become almost invisible in dappled shade. This is also true of the Green-veined White (opposite top) which, in this case, is blending in beautifully with a bramble. Although the caterpillar feeds on Cruciferae, *amongst other plants, it is not normally a pest of the vegetable garden like the Large and Small White.*

Peacock *Inachis io*

We have three skippers – Small, Large and Essex. The Essex and Small Skippers are very similiar. They are not easily seen flitting in and out of grasses that are the main source of food for the larvae. The Essex Skipper may be distinguished by a black spot on the tip of the underside of the antennae. All three species fly around the same time, from May to August, and I have seen them frequently in the glade on the south side of Shaw Wood, the nearby rides and the 'orchid meadow' on the lawn. Numbers vary greatly from year to year: in 2003, when most other species were abundant, I rarely saw them.

Red Admiral *Vanessa atalanta*

Essex Skipper *Thymelicus lineola*

Moths

Common name	Scientific name	Food plants/comments
Alder Moth	*Acronicta alni*	Oak, Alder, Blackthorn
Angle Shades	*Phlogophora meticulosa*	Wide range of plants. See illustration p161.
August Thorn	*Ennomos quercinaria*	Birch, Hawthorn, Oak
Barred Red	*Hylaea fasciaria*	Scots Pine, Douglas Fir and other conifers
Barred Yellow	*Cidaria fulvata*	Wild Roses
Beaded Chestnut	*Agrochola lychnidis*	Grasses
Beautiful Brocade	*Lacanobia contigua*	Birch, Bracken, Dock, Oak
Beautiful Golden Y	*Autographa pulchrina*	Variety of woodland and garden flowers
Beautiful Snout	*Hypena crassalis*	Confined largely to Southern England
Black Arches	*Lymantria monacha*	Pine, Oak, Birch. See illustration p152.
Blood-vein	*Timandra griseata*	Docks, Sorrell
Blotched Emerald	*Comibaena bajularia*	Oak
Bright-line Brown-eye	*Lacanobia oleracea*	Various plants including tomatoes
Brimstone Moth	*Opisthograptis luteolata*	Blackthorn, Hawthorn, Rowan
Brindled Green	*Dryobotodes eremita*	Hawthorn, Oak
Brindled White Spot	*Ectropis extersaria*	Woodland plants
Broad-bordered Yellow Underwing	*Noctua fimbriata*	Blackthorn, Dock and others
Brown China Mark	*Elophila nymphaeata*	Pyralid – see illustration p158
Brown House Moth	*Hofmannophila pseudodspetriella*	Carpet pest
Buff Arches	*Habrosyne pyritoides*	Brambles
Buff Ermine	*Spilosoma luteum*	Various wild and garden plants
Buff Footman	*Eilema deplana*	Algae, Hawthorn, Yew
Buff Tip	*Phalera bucephala*	Oak, Lime, Willow. See illustration p153
Burnished Brass	*Diachrysia chrysitis*	Nettle and others
Canary-shouldered Thorn	*Ennomos alniaria*	Birch, Alder, Lime
Carnation Tortrix	*Cacoecimorpha prombana*	Often a pest on flowers and vegetables
Centre-barred Sallow	*Atethmia centrago*	Ash
Cinnabar	*Tyria jacobaeae*	See illustration p154
Clouded Border	*Lomaspilis marginata*	Sallow, Poplar, Hazel
Clouded Buff	*Diracrisia sannio*	Dandelion, Dock
Common Carpet	*Epirrhoe alternata*	Cleavers, Bedstraw
Common Emerald	*Hemithea aestivaria*	Hawthorn, Birch, Oak, Blackthorn
Common Marbled Carpet	*Chloroclysta truncata*	Birch, Hawthorn, Bramble
Common Quaker	*Orthosia stablilis*	Oak, Sallow
Common Rustic	*Mesapamea secalis*	Creeping Soft-grass, Hairy Wood-rush
Common Wainscot	*Mythimna pallens*	Grasses
Common White Wave	*Cabera pusaria*	Alder, Birch and others
Convolvulus Hawk-moth	*Agrius convolvuli*	Bindweed
Copper Underwing	*Amphipyra pyramidea*	Ash, Honeysuckle, Privet
Coxcomb Prominent	*Ptilodon capucina*	Hazel, Birch etc.
Dark Spectacle	*Abrostola triplasia*	Nettle (not common)
Dark Swordgrass	*Agrostis ipsilon*	Immigrant: wild and cultivated plants
Dingy Footman	*Eilema griseola*	Lichens
Dotted Border	*Agriopis marginaria*	Birch, Oak, Hawthorn
Drinker	*Philadonia potatoria*	Grasses
Dun-bar	*Cosmia trapezina*	Oak, Hazel, Birch, Hawthorn
Dusky Thorn	*Ennomos fuscantaria*	Ash
Early Moth	*Theria primaria*	Hawthorn, Blackthorn
Elephant Hawk-moth	*Deilephila elpenor*	See illustration p155
Engrailed	*Ectropis bistortata*	Oak, Broom, Buckthorn
Eyed Hawk-moth	*Smerinthus ocellata*	Willow, Sallow
Feathered Gothic	*Tholera decimalis*	Grasses
Feathered Thorn	*Colotois pennaria*	Oak, Birch, Sallow, Hawthorn
Flame Carpet	*Xanthorhoe designata*	*Cruciferae*
Flame Shoulder	*Ochropleura plecta*	Dock, Plantain Groundsel
Foxglove Pug	*Eupithecia pulchellata*	Common Toadflax
Frosted Orange	*Gortyna flavago*	Burdock, Foxglove, Thistles – see illustration p160
Garden Carpet	*Xanthorhoe fluctuata*	*Cruciferae*
Garden Pebble	*Evergestis forficalis*	Pyralid – *Cruciferae*, possible vegetable pest
Gold Triangle	*Hypsopygia costalis*	Pyralid – Hay and in squirrels' dreys
Grass Emerald	*Pseudoterpna pruinata*	Broom

Common name	Scientific name	Food plants/comments
Green Carpet	*Colostygia pectinataria*	Bedstraw
Green Silver-lines	*Pseudoips fagana*	Beech, Oak, Birch, Hazel. See illustration p159.
Grey Pine Carpet	*Thera obeliscata*	Conifers
Heart and Dart	*Agrotis excamationis*	Garden plants
Herald Moth	*Scoliopteryx libatrix*	Willow, Poplar, Rowan. See illustration p156.
Hoary Footman	*Eilema griseola*	Unusual immigrant: larvae feed on lichens
Humming-bird Hawk-moth	*Macroglossum stellatarum*	Only spotted twice in 1980s on Wallflower. Migrant.
Ingrailed Clay	*Diarsia mendica*	Bramble, Hawthorn, Sallow
Iron Prominent	*Notodonta dromedarius*	Alder, Birch
July Highflyer	*Hydriomena furcata*	Hazel, Sallow
Knot Grass	*Aconicta rumicis*	Dock, Bramble, Thistles
Lackey	*Malacosoma neustria*	Alder, Birch, also ornamentals
Large Emerald	*Geometra papilionaria*	Beech, Hazel
Large Yellow Underwing	*Noctua pronuba*	Grasses and other plants
Latticed Heath	*Semiothisa clathrata*	Clover, Trefoil
Leopard Moth	*Zeuzera pyrina*	Sallow, Ash, Cherry and others
Lesser Broad-bordered Yellow Underwing	*Noctua janthe*	Blackthorn, Dock
Light Emerald	*Compaea margaritata*	Oak, Hawthorn, Birch
Lobster Moth	*Stauropus fagi*	Beech, Birch, Hazel
Lunar Underwing	*Omphaloscelis lunosa*	Various grasses
Magpie	*Abraxas grossulariata*	Blackthorn, Spindle, Hazel
Merveille du Jour	*Dichonia aprillina*	Buds and leaves of Oak. See illustration p161.
Mother of Pearl	*Pleuroptya ruralis*	Pyralid – Nettles
Mother Shipton	*Calistege mi*	Grasses
Mottled Beauty	*Alcis repandata*	Birch, Hawthorn etc.
Mottled Umber	*Erannis defoliaria*	Many trees and shrubs
Muslin Moth	*Diaphora mendica*	Dock, Chickweed, Dandelion
Nut-tree Tussock	*Colocasia coryli*	Beech, Hornbeam, Hazel
Oak Hook-tip	*Drepana binaria*	Oak
Old Lady	*Mormo maura*	Birch, Blackthorn, Spindle
Orange Swift	*Hepialus sylvina*	Bracken, Dandelion, Dock
Peach Blossom	*Thyatira batis*	Brambles. See illustration p157.
Peacock	*Semiothisa notata*	Birch
Pebble Prominent	*Eligmodonta ziczac*	Willow, Poplar etc.
Peppered Moth	*Biston betularia*	Deciduous trees. See illustration p157.
Pine Beauty	*Panolis flammea*	Pine
Pine Carpet	*Thera firmata*	Pine
Pine Hawk-moth	*Hyloicus pinastri*	Scots Pine
Pink-barred Sallow	*Xanthia togata*	Sallow catkins
Poplar Grey	*Acronicta megacephala*	Poplar, Aspen
Poplar Hawk-moth	*Laothoe populi*	Aspen, Willow. See illustration p158.
Poplar Kitten	*Furcula bifida*	Poplar, Aspen
Purple Thorn	*Selenia tetralunaria*	Birch, Oak, Alder etc.
Red Twin-spot Carpet	*Xanthorhoe spadicearia*	Lower plants
Riband Wave	*Idaea aversata*	Chickweed, Knotgrass
Rosy Footman	*Miltochrista miniata*	Lichens. See illustration p160.
Rosy Rustic	*Hydraecia micacea*	Larvae feeds on roots: Dock, Burdock,Plantain
Rush Veneer	*Nomophila noctuella*	Pyralid – Grasses and Clover
Sallow	*Xanthia icteritia*	Sallow catkins and leaves
Satellite	*Eupsilis transversa*	Oak, Beech, Elm. Likes rotting apples.
Scalloped Hazel	*Odontopera bidentata*	Oak, Birch, Hawthorn and others
Scalloped Hook-tip	*Falcaria lacertinaria*	Birch, Alder
Scalloped Oak	*Crocallis elinguaria*	See illustration p154.
Scorched Wing	*Plagodis dolabraria*	Birch, Oak, Sallow
Setaceous Hebrew Character	*Xestia c-nigrum*	Oak, Chickweed, Groundsel
Shoulder-striped Wainscot	*Mythimna comma*	Grasses
Silver Y	*Autographa gamma*	Migrant – vegetables and flowers
Six-striped Rustic	*Xestia sexstrigata*	Various plants
Small Elephant Hawk-moth	*Deilephila porcellus*	Bedstraw
Small Engrailed	*Ectropis crepuscularia*	Birch, Sallow
Small Fan-footed Wave	*Idaea biselata*	Dandelion, Bramble, Knotgrass
Small Phoenix	*Ecliptopera silaceata*	Willowherbs
Small Yellow Wave	*Hydrelia flammeolaria*	Alder
Smoky Wainscot	*Mythimna impura*	Grasses
Snout	*Hypena proboscidalis*	Nettles

Common name	Scientific name	Food plants/comments
Speckled Yellow	*Pseudopanthera macularia*	Wood Sage, Yellow Archangel, Dead-nettle. See p159.
Spectacle	*Abrostola tripartita*	Nettles
Spruce Carpet	*Thera britannica*	Douglas Fir, Norway Spruce
Square-spot Rustic	*Xestia xanthographa*	Grasses
Straw Dot	*Rivula sericealis*	Mainly grasses
Straw Underwing	*Thalpophila matura*	Grasses
Svensson's Copper Underwing	*Amphipyra berbera*	Oak, Lime, Hornbeam
Swallow Prominent	*Pheosia tremula*	Poplar, Willow
Tawny-barred Angle	*Semiothisa liturata*	Scots Pine, Norway Spruce
Treble Lines	*Charanyca trigrammica*	Common Knapweed
Vapourer	*Orgyia antiqua*	Most deciduous trees. See illustration p159.
White Ermine	*Spilosoma lubricipeda*	Various plants
White Plume	*Pterophorus pentadactyla*	Bindweed. See illustration p155.
Willow Beauty	*Peribatodes rhomboidaria*	Various trees
The following are Pyralid Moths without common names		
	Agriphila tristella	Tall grasses
	Chrysoteuchia culmella	Stems of grasses
	Endotricha flammealis	Bird's-foot-trefoil and decaying leaves
	Eudonia truncicolella	Mosses

Black Arches *Lymantria monacha*

*This moth is confined mainly to southern parts of England where it feeds largely on Oak. In some continental forests it can become a pest like its cousin the Gipsy Moth (*Lymantria dispar*). There is no such evidence here which, I believe, is due to the fact that we no longer have any mono-silviculture. It is often the absence of a wide biodiversity and, therefore, no natural balance, that enables a single species to turn into a pest.*

Buff Tip *Phalera bucephala*

What a master of disguise! If this moth places himself on a broken twig, you would have difficulty in spotting him, as would all the other predators (birds, bats and insects) in search of a meal. His main defence is obviously to sit like this as still as possible – he never budged once whilst having his photograph taken! His underwings are almost white, which would make him visible and vulnerable.

Scalloped Oak *Crocallis elinguaria*

This moth flits around Oak, Birch, Hazel and many other deciduous trees, which are the caterpillar's source of food. Like the Buff Tip, it is not easily spotted when at rest.

Cinnabar Moth *Tyria jacobaeae*

A lot of unnecessary fuss is often made about Common Ragwort, which is the major source of food for this moth. We have plenty of Ragwort, but my ponies never touch it – the answer seems to be never to cut for hay if this normally very conscipuous plant is present. The moth itself, which has mainly black forewings and crimson hindwings, usually hides in the undergrowth during the day.

Elephant Hawk-moth *Deilephila elpenor*

Rosebay Willowherb is one of the Elephant Hawk-moth's favourite plants and the adult is well-equipped to hide itself among the flowers and leaves. However, the name comes from the caterpillar that sometimes stretches out the first few segments of its body, giving the illusion that it has a large trunk.

White Plume *Pterophorus pentadactyla*

*This does not belong to the macro group of moths, but as it is so striking, I could not resist including it. It is, in fact, a member of the plume moth family (*Pterophoridae*). It was not caught in the moth trap, but turned up one day in the conservatory. I wish there were more of them, as the larvae feed on species of convolvulus that abound in the vegetable garden.*

Herald Moth *Scoliopteryx libatrix*

This is another moth with a masterly disguise. It flies well into the winter, and here it is sitting on oak leaves. It was not caught in the moth trap but flew into the house in December.

Peppered Moth *Biston betularia*

This is another moth that, once settled, sits perfectly still. It is also the moth that sprang to worldwide fame when it was discovered that it could demonstrate evolution in a very short time. Those typically-coloured specimens that got caught up in the industrial revolution around the start of the 20th century quickly became preferentially eaten by birds when resting on a dark background, so that the proportion of blacker, sootier examples in the population increased.

Peach Blossom *Thyatira batis*

Some moths have wonderfully evocative names, and with imagination one could compare this to blossoms. It never sat still, and either flew off or sat with quivering wings as in this picture.

Poplar Hawk-moth *Laothoe populi*

This handsome moth is a good photographic model! It sits quite still wherever you care to put it, but if you attempt to open its wings to display the conspicuous red-marked underwing, it becomes agitated and usually flies off. Undoubtedly it does not want to be seen by any potential predator. It will also walk up your hand or finger, clinging on so fiercely it produces a prickling sensation.

Brown China Mark *Elophila nymphaeata*

This pyralid moth, with a wingspan of two and a half centimetres, would indeed make an attractive pattern on a porcelain plate, but I have no idea whether the name-giver had this in mind. The larva feeds on water plants, including Bog Pondweed and Bur-reed, both of which grow in Pope's Wood pond.

Speckled Yellow *Pseudopanthera macularia*

This moth usually flies by day. Its colouring is a perfect camouflage in the dappled woodland sunlight.

Green Silver-lines *Pseudoips fagana*

This moth is usually found in the south of England, and the larvae feed on very common trees here: Oak, Birch, Beech and Hazel.

Vapourer *Orgyia antiqua*

The moth of this caterpillar is far less flamboyant, although the male does have a pair of spectacular antennae. The female is wingless and looks like a grub or beetle. The larva, for some reason, reminds me of a nineteenth-century steam train: often brightly coloured, although never, I think, with four chimney funnels!

Frosted Orange *Gortyna flavago*

I find this a stunning moth and it has a name to go with its appearance. It has a wingspan of about 40mm. The larva feeds on many plants including thistles, burdocks and foxgloves.

Rosy Footman *Miltochrista miniata*

I wonder where some moths get their extravagant sounding names? There are many 'footmen' and most of them can be found in this part of England. They include the Dingy, Muslin, Common, Red-necked, Four-dotted, Orange, Pygmy and Scarce. Apparently some, when resting in a vertical position, resemble dressed-up footmen. The one pictured here is nothing like that, but at least it has the right colour. Many of the larvae, including this one, feed on lichens, of which there are many here.

Merveille du Jour *Dichonia aprillina*

Translated literally, this moth is 'Wonder of the Day'. Certainly the green, black and white markings make it wondrously beautiful! Its Latin name seems to imply that the adult emerges in April, which it certainly does not. The name has always been a mystery. The larvae feed on Oak buds and leaves, after which they pupate in earth amongst tree roots.

Angle Shades
Phlogophora meticulosa

This is a distinctive and beautiful moth; everything about it is designed to give it camouflage against birds and other predators. Angle Shades also come in shades of green and orange. The skin of the larvae is translucent and they feed on a number of plants (including nettles, docks, brambles, birches, oaks) and, depending on the food plant, the caterpillar takes on those colours.

Dragonflies and Miscellaneous

I mentioned that Tim Harris persuaded me at the last minute to include dragonflies as I had already, contrary to my original intention, included some other insects. He was quite right, as dragonflies are astonishingly beautiful, but it was impossible for him to come up with a full list in a few hours at the end of summer/beginning of autumn. I know there are many more, and I was able to add two which I had photographed some years back.

I persuaded the publishers to allow me a few extra pages, so I reinstated a miscellaneous section, which I had discarded for reasons of space. In this I wanted to show some of the more unusual sights that I had come across over the years and had managed to photograph. They are pictured on the next four pages.

Dragonflies

Common name	Scientific name	Food plants/comments
Azure Damselfly	*Coenagrion puella*	Flies late May–August. Lays eggs in submerged pond vegetation.
Beautiful Demoiselle	*Calopteryx virgo*	Flies late May–August. Prefers pebble-bottomed or sandy-bottomed streams.
Brown Hawker	*Aeshna grandis*	Flies late July–September. Tolerates some pollution in ponds and streams.
Common Blue Damselfly	*Enallagma cyathigerum*	Flies late May–September. Lays eggs in submerged pond vegetation
Emperor Dragonfly	*Anex imperator*	Flies June–early August
Golden-ringed Dragonfly	*Cordulegaster boltonii*	Flies June–August. See illustration above.
Large Red Damselfly	*Pyrrhosoma nymphula*	Flies late May–early August. Lays eggs on submerged vegetation or on underside of floating leaves.
Migrant Hawker	*Aeshna mixta*	Flies August and September. Eggs often laid on bulrush.
Southern Hawker	*Aeshna cyanea*	Flies late July–October. Female sometimes oviposits on woodland rides.

Left
Golden-ringed Dragonfly
Cordulegaster boltonii

We seemed to have many Golden-ringed Dragonflies. The one pictured opposite is extraordinarily handsome, but even more amazing was the fact that he sat unmoving on a Hazel twig on the border of Pope's Wood and Butler's Wood whilst I was summoned by Tim to come and take a photograph. He must have sat there twenty minutes or more! The Golden-ringed is commonly found in the west of England, Wales and the North. Here, we seem to be on the fringes of its range. It prefers clean, fast-flowing rivers, which we do not have here, but the tiny streams in Pope's Wood (at least those emanating from our constantly overflowing spring) are clear and clean, with plenty of silt and overhanging grasses and sedges in which the nymphs can grow to maturity. It takes four or five years for the adult to finally emerge. The adults' food consists of various insects, while the larvae seem to be voracious feeders, managing to consume even small fish.

Beautiful Demoiselle *Calopteryx virgo*

Glow-worm *Lampyris noctiluca*

The Glow-worm is not a worm but a beetle. The female larva, pictured here, is the one that emits the 'glow' – looking a bit like a lighted cigarette. She produces this light in order to attract flying males who do not glow. The light comes from the underside of the female's body where there is a chemical called luciferin. *She is able to switch this light on and off at will – she will normally switch off if you pick her up. Glow-worms feed on tiny garden snails. There is a gap of two or three years between mating and the emergence of the adult.*

I have found them only in the garden, normally round the edges of the shrubbery which is seldom cut or weeded. They probably also exist on the wide rides in the woods. There must be something about summer thunder and lightning that stimulates them to glow en masse. *I remember on several stormy nights looking out of the window and seeing the fringes of the lawn lit up with literally dozens of tiny lights – an unforgettable sight.*

Here is rather a nice little ditty about Glow-worms, author anon.

I wish I was a glow worm
A glow worm's never glum
'Cos how can you be grumpy
When the sun shines out your bum?

Ladybird, pupae and larvae

*Ladybirds have a good, friendly image, unlike wasps and hornets; probably enhanced by the fact that ladybirds consume garden pests, such as aphids. There are twenty-five species of ladybird in Britain and the one above is an Orange Ladybird (*Halyzia sedecimguttata*), found less frequently than most. Unlike its cousins, it does not feed on insects and lives in dead wood.*

*The pictures above and left are probably the larva and pupa of the common red Seven-spot Ladybird (*Coccinella 7-punctata*). You often see this ladybird but you may not recognise the pupae and larvae. The larva is a voracious predator of insects, including aphids.*

Wasp's nest

*The Common Wasp (*Vespula vulgaris*) and the Hornet (*Vespa crabro*) opposite, have a really bad image. Most people's instinct is to swat or squash them immediately, but they will not sting unless frightened or provoked. They usually come into houses either by mistake or because they have sensed the presence of something sweet like jam or honey. The adults will feed on nectar, but they will also catch all kinds of insects to take back to their larvae, which live in beautifully fashioned nests (made out of shredded wood) like the one opposite. Nests are often underground, or in lofts and sheds, but not normally outside, hanging on a shrub like this one.*

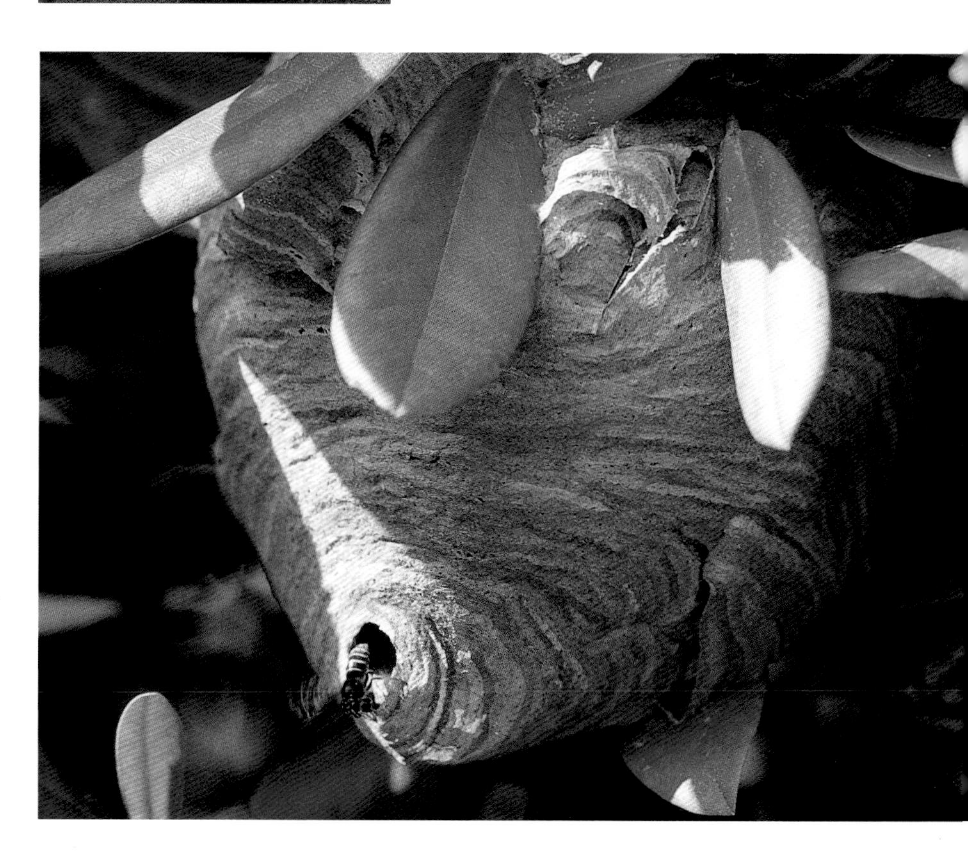

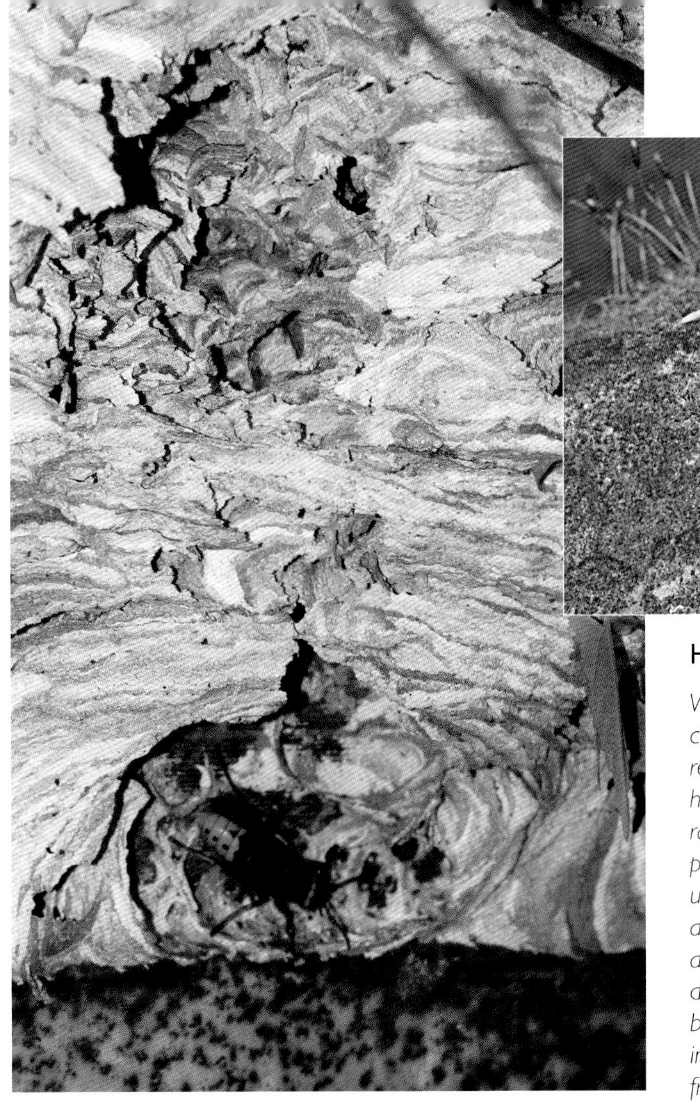

Hornet's nest, Hornet and (below) bee swarm

When someone finds a swarm of bees (below) they will usually contact the local Bee Keepers' Association to come and remove it. This swarm was unusually close to the ground, hanging from a Cotoneaster bush. Hornets' nests are much rarer, and I dare say if anyone finds one, they will contact the pest exterminator! However, like wasps, they will not attack unless they feel threatened. The inhabitants of this nest allowed me to carefully cut away branches obscuring the view and let me get quite close to photograph them. They came dangerously near and seemed to look me straight in the eye, but left me unmolested. I rarely found hornets in the past, but in the last two years (2002 and 2003) they became quite frequent due, perhaps, to climate change?

Leech

There are about twenty species of leech (Phylum Annelida) *in Britain. They are related to earthworms and include the now rare medicinal leech. The first thing that Barry Kemp fished out of the pond in Shaw Wood (when trying to survey the amphibians) was this leech – it immediately drew blood!*

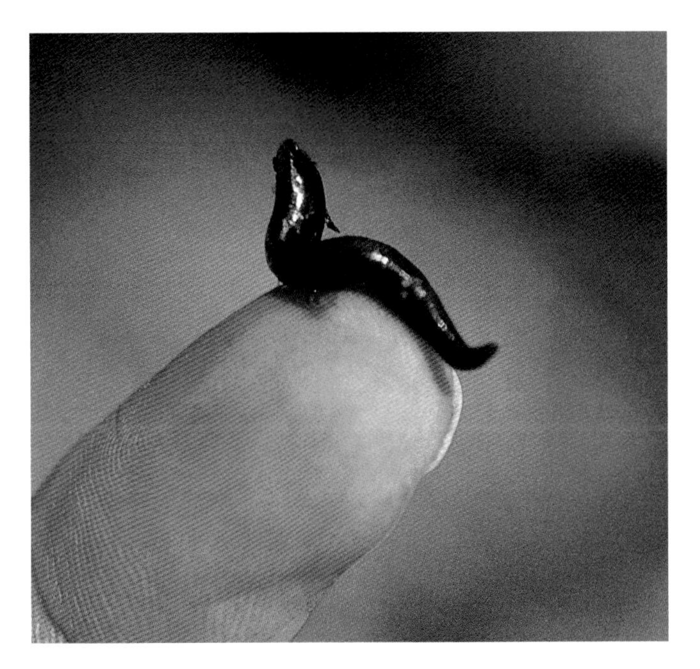

CONCLUSION

This book started with the idea of creating no more than a glorified scrapbook of the woods for my family. It has ended up as something far more substantial with, I hope, a more serious intent. I wonder, if you have got to the end of this book, what, if any, conclusions you have drawn? For me it has certainly reinforced my belief that this small piece of land is indeed rich in biodiversity. Of course, I am prejudiced, but I would go as far as claiming there are not many areas of this size that can equal it.

However, one can draw a much less positive conclusion: it is never going to be easy to retrieve or create diverse ecosystems. It is one thing to latch on to ancient sites (how many of those are left?), but quite another to recreate them in almost total isolation. Our modern life-styles and food production methods are responsible for the steady erosion of our biodiversity, and it will take centuries to recover what has been lost in just a few decades.

There are some examples in our forty acres: Streake's Wood, adjacent to ancient woodland, has taken nearly a century to achieve any sort of biodiversity. Even the ancient woodland sites that were under mono-silviculture will take many decades to fully recover despite the fact that the short-term transformation seemed dramatic. It remains to be seen what will happen in Morgan's Strip. Imagine planting forty acres of woodland isolated amidst intensive farmland!

I have always been convinced that the natural world depends totally on the soil, the micro-organisms and creatures that inhabit it. It is there that one must start with the creation or recovery of any ecosystems. Everything I have learnt during the course of writing this book has reinforced this belief. In fact, had I been a scientist intent on demonstrating the biodiversity of this small plot, I would probably have started with the soil, progressing to beetles and insects right up the whole food chain, instead of commencing with the trees and working my way down! Maybe someone will undertake such a daunting project in the future?

The writing of the book has been a wonderful experience and an enormous learning curve for me. Suddenly, one becomes aware of 'things' that have been on the peripherals of one's vision: suddenly they become intriguing, living organisms that cry out to be preserved and studied in greater depth. It would not take much to become hooked on moths or mosses, or indeed anything else!

Having had people in and around the woods who have devoted themselves to one or other of these disciplines has also been an exceptional experience, and I admire the commitment to their particular subjects. As I have already said, without them this book would not have been possible. Maybe one day we will be able to bring the surveys up-to-date and augment those that I know are far from complete.

Finally, I hope that readers may have gained just a little and may in future look more closely at our awe-inspiring biodiversity. More importantly, I hope it will persuade some, if not actively to contribute to the increase of our biodiversity, at least to take positive steps to preserve what still remains.

Postscript
Since completing this book, I am happy to report that the Sussex Black Poplar Group has agreed to come and plant some true Sussex Black Poplars in the autumn of 2004. These will be planted in Pope's Wood, largely to replace the cloned Poplars that have either been felled or ring-barked. Also, the Sussex Otters and Rivers Partnership will build an otter holt close to the woods. Surely there is an optimistic outlook for the biodiversity of these small woods.

Index of Scientific and Common Names

Scientific names are in italic and common names are in roman type. Figures in bold denote illustrations

Flora

Plants

Fungi, Mosses and Lichens

Fauna

Pope's Wood 'Doodlebug' pond, Spring 2004

General Index

Names of woods in bold, publications in italics

The new ride in Butler's Wood, Spring 2004